Learning Geospatial Ar with Python

with Python
Third Edition

Understand GIS fundamentals and perform remote sensing
data analysis using Python 3.7

Joel Lawhead

BIRMINGHAM - MUMBAI

Learning Geospatial Analysis with Python
Third Edition

Commissioning Editor: Richa Tripathi
Acquisition Editor: Denim Pinto
Content Development Editor: Ruvika Rao
Senior Editor: Afshaan Khan
Technical Editor: Gaurav Gala
Copy Editor: Safis Editing
Project Coordinator: Prajakta Naik
Proofreader: Safis Editing
Indexer: Tejal Daruwale Soni
Production Designer: Jyoti Chauhan

First published: October 2013
Second edition: December 2015
Third edition: September 2019

Production reference: 1270919

Published by Packt Publishing Ltd.
Livery Place
35 Livery Street
Birmingham
B3 2PB, UK.

ISBN 978-1-78995-927-7

www.packt.com

To my wife, Julie, my coauthor of the incredible 22-year work of time, geography, and love that is our life.

To my children, Lauren, Will, Lillie, and Lainie, I love you all and hope each one of you hosts your own podcast one day that I can listen to on car trips.

To NVision Solutions, the team that first introduced me to geospatial technology almost 20 years ago and continues to create a challenging environment focused on continual learning and growth. Thank you Lalet, Craig, Amy, April, Joel "JJ" Herr, Mark, and Darren for all of the fun, interesting projects, and management meetings.

To the elite runners of the very prestigious Cedar Point Running Club, especially Robin, Leanne, and Mandy, for helping to collect sample GPS data for this book, and breaking up the long hours of coding, writing, and editing with brain-refreshing, soul-rejuvenating runs.

Subscribe to our online digital library for full access to over 7,000 books and videos, as well as industry leading tools to help you plan your personal development and advance your career. For more information, please visit our website.

Why subscribe?

- Spend less time learning and more time coding with practical eBooks and Videos from over 4,000 industry professionals

- Improve your learning with Skill Plans built especially for you

- Get a free eBook or video every month

- Fully searchable for easy access to vital information

- Copy and paste, print, and bookmark content

Did you know that Packt offers eBook versions of every book published, with PDF and ePub files available? You can upgrade to the eBook version at www.packt.com and as a print book customer, you are entitled to a discount on the eBook copy. Get in touch with us at customercare@packtpub.com for more details.

At www.packt.com, you can also read a collection of free technical articles, sign up for a range of free newsletters, and receive exclusive discounts and offers on Packt books and eBooks.

Contributors

About the author

Joel Lawhead is a PMI-certified Project Management Professional, a certified GIS Professional, and the Chief Information Officer of NVision Solutions Inc., an award-winning firm specializing in geospatial technology integration and sensor engineering for NASA, FEMA, NOAA, the US Navy, and many other commercial and non-profit organizations. Joel began using Python in 1997 and started combining it with geospatial software development in 2000. He has authored multiple editions of *Learning Geospatial Analysis with Python* and *QGIS Python Programming Cookbook*, both from Packt. He is also the developer of the open source Python Shapefile Library (PyShp) and maintains a geospatial technical blog, *GeospatialPython*, and Twitter feed, `@SpatialPython`.

About the reviewer

Athanasios Tom Kralidis is a Senior Systems Scientist at the Meteorological Service of Canada, where he provides geospatial technical and architectural leadership in support of weather, climate and water data delivery. He is an active, long-time contributor to international standards and activities at the Open Geospatial Consortium and the World Meteorological Organization, as well as free and open source geospatial software and the Open Source Geospatial Foundation.

Packt is searching for authors like you

If you're interested in becoming an author for Packt, please visit `authors.packtpub.com` and apply today. We have worked with thousands of developers and tech professionals, just like you, to help them share their insight with the global tech community. You can make a general application, apply for a specific hot topic that we are recruiting an author for, or submit your own idea.

Table of Contents

Preface

The book starts by giving you the background of geospatial analysis, and then offers a flow of the techniques and technology used and splits the field into its component specialty areas, such as **Geographic Information Systems** (**GIS**), remote sensing, elevation data, advanced modeling, and real-time data. The focus of the book is to lay a strong foundation in using the powerful Python language and framework to approach geospatial analysis effectively. In doing so, we'll focus on using pure Python as well as certain Python tools and APIs, and using generic algorithms. The readers will be able to analyze various forms of geospatial data, learn about real-time data tracking, and see how to apply what they learn to interesting scenarios.

While many third-party geospatial libraries are used throughout the examples, a special effort will be made by us to use pure Python, with no dependencies, whenever possible. This focus on pure Python 3 examples is what will set this book apart from nearly all other resources in this field. We will also go through some popular libraries that weren't in the previous version of the book.

Who this book is for

This book is for anyone who wants to understand digital mapping and analysis and who uses Python or any other scripting language for the automation or crunching of data manually. This book primarily targets Python developers, researchers, and analysts who want to perform geospatial modeling and GIS analysis with Python.

What this book covers

Chapter 1, *Learning about Geospatial Analysis with Python*, introduces geospatial analysis as a way of answering questions about our world. The differences between GIS and remote sensing are explained. Common geospatial analysis processes are demonstrated using illustrations, basic formulas, pseudo code, and Python.

Chapter 2, *Learning Geospatial Data*, explains the major categories of data and several newer formats that are becoming more and more common. Geospatial data comes in many forms. The most challenging part of geospatial analysis is acquiring the data that you need and preparing it for analysis. Familiarity with these data types is essential to understanding geospatial analysis.

Chapter 3, *The Geospatial Technology Landscape*, tells you about the geospatial technology ecosystem, which consists of thousands of software libraries and packages. This vast array of choices is overwhelming for newcomers to geospatial analysis. The secret to learning geospatial analysis quickly is understanding the handful of libraries and packages that really matter. Most other software is derived from these critical packages. Understanding the hierarchy of geospatial software and how it's used allows you to quickly comprehend and evaluate any geospatial tool.

Chapter 4, *Geospatial Python Toolbox*, introduces software and libraries that form the basis of the book and are used throughout. Python's role in the geospatial industry is explored: the GIS scripting language, the mash-up glue language, and the full-blown programming language. Code examples are used to teach data editing concepts, and many of the basic geospatial concepts in Chapter 1, *Learning about Geospatial Analysis with Python*, are also demonstrated in this chapter.

Chapter 5, *Python and Geographic Information Systems*, teaches you about simple yet practical Python GIS geospatial products using processes that can be applied to a variety of problems.

Chapter 6, *Python and Remote Sensing*, shows you how to work with remote sensing geospatial data. Remote sensing includes some of the most complex and least-documented geospatial operations. This chapter will build a solid core for you and demystify remote sensing using Python.

Chapter 7, *Python and Elevation Data*, demonstrates the most common uses of elevation data and how to work with its unique properties. Elevation data deserves a chapter all on its own. Elevation data can be contained in almost any geospatial format but is used quite differently from other types of geospatial data.

Chapter 8, *Advanced Geospatial Python Modeling*, uses Python to teach you the true power of geospatial technology. Geospatial data editing and processing help us understand the world as it is. The true power of geospatial analysis is modeling. Geospatial models help us predict the future, narrow a vast field of choices down to the best options, and visualize concepts that cannot be directly observed in the natural world.

Chapter 9, *Real-Time Data*, examines the modern phenomenon of geospatial analysis. A wise geospatial analyst once said, "As soon as a map is created it is obsolete." Until recently, by the time you collected data about the Earth, processed it, and created a geospatial product, the world it represented had already changed. But modern geospatial data shatters this notion. Datasets are available over the internet that are up to the minute, or even the second. This data fundamentally changes the way we perform geospatial analysis.

Chapter 10, *Putting It All Together*, combines the skills from the previous chapters step by step to build a generic corporate system to manage customer support requests and field support personnel that could be applied to virtually any organization.

To get the most out of this book

This book assumes you have basic knowledge of the Python programming language. You will require Python (3.7 or higher), a minimum hardware requirement of a 300-MHz processor, 128 MB of RAM, 1.5 GB of available hard disk, and a Windows, Linux, or macOS X operating system.

Download the example code files

You can download the example code files for this book from your account at www.packt.com. If you purchased this book elsewhere, you can visit www.packtpub.com/support and register to have the files emailed directly to you.

You can download the code files by following these steps:

1. Log in or register at www.packt.com.
2. Select the **Support** tab.
3. Click on **Code Downloads**.
4. Enter the name of the book in the **Search** box and follow the onscreen instructions.

Once the file is downloaded, please make sure that you unzip or extract the folder using the latest version of:

- WinRAR/7-Zip for Windows
- Zipeg/iZip/UnRarX for Mac
- 7-Zip/PeaZip for Linux

The code bundle for the book is also hosted on GitHub at https://github.com/PacktPublishing/Learning-Geospatial-Analysis-with-Python-Third-Edition. In case there's an update to the code, it will be updated on the existing GitHub repository.

We also have other code bundles from our rich catalog of books and videos available at https://github.com/PacktPublishing/. Check them out!

Download the color images

We also provide a PDF file that has color images of the screenshots/diagrams used in this book. You can download it here: `https://static.packt-cdn.com/downloads/9781789959277_ColorImages.pdf`.

Conventions used

There are a number of text conventions used throughout this book.

`CodeInText`: Indicates code words in text, database table names, folder names, filenames, file extensions, pathnames, dummy URLs, user input, and Twitter handles. Here is an example: "To demonstrate this, the following example accesses the same file that we just saw but by using `urllib` instead of `ftplib`."

A block of code is set as follows:

```
import ftplib

server = "ftp.ngdc.noaa.gov"
dir = "hazards/DART/20070815_peru"
fileName = "21415_from_20070727_08_55_15_tides.txt"
```

When we wish to draw your attention to a particular part of a code block, the relevant lines or items are set in bold:

```
if (sinSigma == 0):
        distance = 0   # coincident points
        break
    cosSigma = sinU1*sinU2 + cosU1*cosU2*cosLam
    sigma = math.atan2(sinSigma, cosSigma)
    sinAlpha = cosU1 * cosU2 * sinLam / sinSigma
    cosSqAlpha = 1 - sinAlpha**2
```

Any command-line input or output is written as follows:

```
pip install virtualenv
```

Bold: Indicates a new term, an important word, or words that you see onscreen. For example, words in menus or dialog boxes appear in the text like this. Here is an example: "Select **System info** from the **Administration** panel."

Warnings or important notes appear like this.

 Tips and tricks appear like this.

Get in touch

Feedback from our readers is always welcome.

General feedback: If you have questions about any aspect of this book, mention the book title in the subject of your message and email us at customercare@packtpub.com.

Errata: Although we have taken every care to ensure the accuracy of our content, mistakes do happen. If you have found a mistake in this book, we would be grateful if you would report this to us. Please visit www.packtpub.com/support/errata, selecting your book, clicking on the Errata Submission Form link, and entering the details.

Piracy: If you come across any illegal copies of our works in any form on the Internet, we would be grateful if you would provide us with the location address or website name. Please contact us at copyright@packt.com with a link to the material.

If you are interested in becoming an author: If there is a topic that you have expertise in and you are interested in either writing or contributing to a book, please visit authors.packtpub.com.

Reviews

Please leave a review. Once you have read and used this book, why not leave a review on the site that you purchased it from? Potential readers can then see and use your unbiased opinion to make purchase decisions, we at Packt can understand what you think about our products, and our authors can see your feedback on their book. Thank you!

For more information about Packt, please visit packt.com.

Section 1: The History and the Present of the Industry

1

This section starts by demonstrating common geospatial analysis processes using illustrations, basic formulas, simple code, and Python. Building on that, you'll learn how to play with geospatial data—acquiring data and preparing it for various analyses. After that, you'll gain an understanding of the various software packages and libraries used in the geospatial technology ecosystem. At the end of this section, you'll learn how to evaluate any geospatial tool.

This section includes the following chapters:

- Chapter 1, *Learning about Geospatial Analysis with Python*
- Chapter 2, *Learning Geospatial Data*
- Chapter 3, *The Geospatial Technology Landscape*

1
Learning about Geospatial Analysis with Python

Geospatial technology is currently impacting our world since it is changing our knowledge of human history. In this book, we will step through the history of geospatial analysis, which predates computers and even paper maps. Then, we will examine why you might want to learn about and use a programming language as a geospatial analyst as opposed to just using **geographic information system** (**GIS**) applications. This will help us understand the importance of making geospatial analysis as accessible as possible to as many people as possible.

In this chapter, we will be covering the following topics:

- Geospatial analysis and our world
- Dr. Sarah Parcak and archaeology
- Geographic information systems
- Remote sensing concepts
- Elevation data
- Computer-aided drafting
- Geospatial analysis and computer programming
- The importance of geospatial analysis
- Geographic information system concepts
- Common GIS processes
- Common remote sensing processes
- Common raster data concepts
- Creating the simplest possible Python GIS

Yes, you heard that right! We will be building the simplest possible GIS from scratch using Python, right from the start.

Technical requirements

This book assumes that you have some basic knowledge of the Python programming language, basic computer literacy, and at least an awareness of geospatial analysis. This chapter provides a foundation for geospatial analysis, which is needed to attack any subject in the areas of remote sensing and GIS, including the material in all the other chapters of this book.

The examples in this book are based on Python 3.4.3, which you can download here: `https://www.python.org/downloads/release/python-343/`.

Geospatial analysis and our world

In the 1880s, British explorers began applying scientific rigor to excavating ancient cultural sites. The field of archaeology is a frustrating, low, costly, and often dangerous endeavor requiring patience and a good bit of luck. The Earth is remarkably good at keeping secrets and erasing the story of human endeavors. Changing rivers, floods, volcanoes, dust storms, hurricanes, earthquakes, fires, and other events swallow entire cities into the surrounding landscape, and we lose them to the flow of time.

Our knowledge of human history is based on glimpses into ancient cultures through archaeological excavation and the study of sites we have been lucky enough to stumble across through educated guesses or trial and error. There used to be no success in archaeology unless a team excavated a site, found something, and correctly identified it. Predictions on where to look were based on a handful of major factors such as proximity to water that was needed to support agriculture, previously discovered sites, accounts by early explorers, and other broad clues.

In 2007, archeologist **Dr. Sarah Parcak**, from the University of Alabama, Birmingham, began to coax our stubborn Earth into revealing its secrets about where humans have been and what they've done. Since then, her approach has revolutionized the field of archaeology.

In a few short years, Dr. Parcak and her team found traces of 17 pyramids, more than 1,000 tombs, and the footprints of 3,000 ancient settlements in Egypt, including the city grid of the famous lost city of Tanis. She identified significant archaeological sites in Romania, the Nabataean Kingdom, and Tunisia. She located an arena at the well-excavated ancient Roman harbor of Portus, as well as its lighthouse and canal leading to Rome near the Tiber river.

How did she find so many hidden treasures that eluded detection for almost two centuries? She looked at the bigger picture. Dr. Parcak perfected the art of using satellite imagery to locate ancient sites from almost 400 miles above the Earth. Her career happened to coincide with the advent of readily-available, high-resolution satellite imagery that had a 10-inch pixel resolution or less, thereby providing the detail that was needed to detect subtle changes in the landscape, thus indicating ancient sites.

Despite the volume and significance of her finds, locating cultural heritage sites from space requires a tremendous amount of work. Space archaeologists first research old maps and historical accounts. Then, they look at modern digital maps of existing sites. They also look at digital terrain models to locate subtle rises in the land where ancient people would build to avoid floods. Then, they use multispectral imagery, including infrared, which can expose changes in vegetation or soil when processed due to imported stone and other materials buried underground that bubble up to the surface. This discoloration, which is represented by false colors, allows us to differentiate between the bandwidths of sunlight reflected from sites that are completely invisible on the ground, or even from the air, to the naked eye, which suddenly stand out in sharp contrast, showing precise locations on a satellite image.

Ancient cultural sites are often invisible to the naked eye from the ground. For example, the following photograph shows a well-preserved Native American burial mound near Lewiston, Illinois, USA, which has survived for thousands of years due to its location and is easily visible:

However, in areas with harsher weather conditions, sites can be partially destroyed, and so they are difficult to find. The following photograph shows an area of marsh in Louisiana, which is full of ancient Native American burial mounds that have eroded over the centuries and are now nearly impossible to detect without satellite images:

The following processed satellite image, from NASA scientist Dr. Marco Giardino, is in the same marsh area as the previous photograph and shows the remains of four distinct burial mounds that aren't visible from the ground. Even though this site is hundreds of years old, the vegetation species and their health are different compared to the surrounding marsh. Although archaeologists researched dozens of similar sites in the area, this project was the first to determine that the mound builders often used a pattern of placing the mounds in the four cardinal directions (north, south, west, east), which is highly visible from space but difficult to realize on the ground:

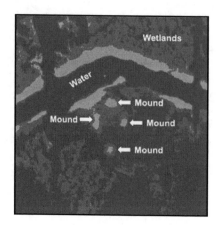

As quick as space archaeologists are at locating ancient sites, they now find themselves battling more than geological and meteorological elements. Looting has always been a threat to archaeology, but due to warfare and black market artifacts, it has become even more of a problem. Modern construction can also destroy valuable sites. However, determined archaeologists are using the same technology they used to find the sites mentioned in order to detect looting or construction threats. Once they find evidence of a threat, they notify the government so that they can intervene. The following image shows evidence of looting at the Roman site of Dura Europos in eastern Syria. The circled areas contain holes that were dug by looters:

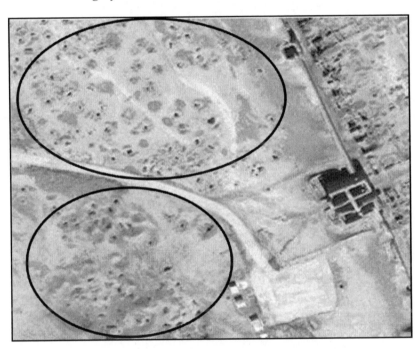

In addition to satellite image processing and visual interpretation, space archaeologists also use geographic information system mapping techniques to mark or digitize sites, overlay modern roads and city footprints, create labeled maps, and much more. The exciting new field of space archaeology is one of the latest of many applications of geospatial analysis we will cover in this book.

Beyond archaeology: Geospatial analysis can be found in almost every industry, including real estate, oil and gas, agriculture, defense, disaster management, health, transportation, and oceanography, to name a few. For a good overview of how geospatial analysis is used in dozens of different industries, go to `https://www.esri.com/what-is-gis/who-uses-gis`.

History of geospatial analysis

Geospatial analysis can be traced back to as far as 15,000 years ago, to the Lascaux cave in southwestern France. In this cave, Paleolithic artists painted commonly hunted animals and what many experts believe are astronomical star maps for either religious ceremonies or potentially even migration patterns of prey. Though crude, these paintings demonstrate an ancient example of humans creating abstract models of the world around them and correlating spatial-temporal features to find relationships. The following photograph shows one of the paintings, with an overlay illustrating the star maps:

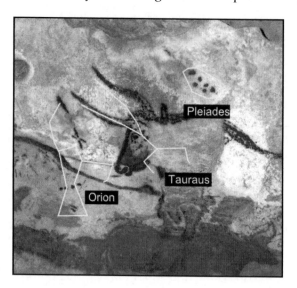

Over the centuries, the art of cartography and the science of land surveying have developed, but it wasn't until the 1800s that significant advances in geographic analysis emerged. Deadly cholera outbreaks in Europe between 1830 and 1860 led geographers in Paris and London to use geographic analysis for epidemiological studies.

In 1832, Charles Picquet used different halftone shades of gray to represent the deaths per thousand citizens in the 48 districts of Paris as part of a report on the cholera outbreak. In 1854, Dr. John Snow expanded on this method by tracking a cholera outbreak in London as it occurred. By placing a point on a map of the city each time a fatality was diagnosed, he was able to analyze the clustering of cholera cases. Snow traced the disease to a single water pump and prevented further cases. The following map has three layers with streets, an **X** for each pump, and dots for each cholera death:

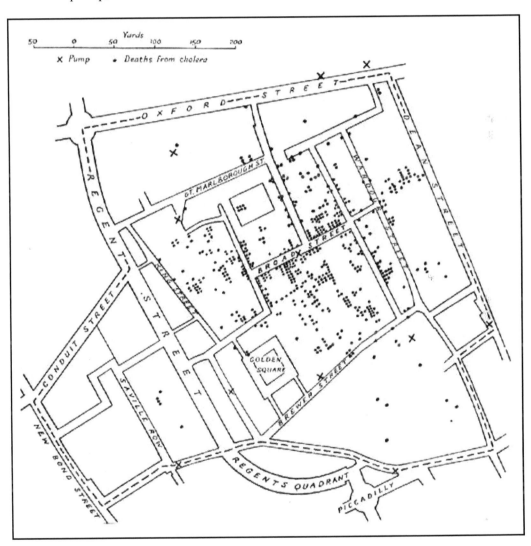

Geospatial analysis wasn't just used for the war on diseases. For centuries, generals and historians have used maps to understand human warfare. A retired French engineer named Charles Minard produced some of the most sophisticated infographics that were ever drawn between 1850 and 1870. The term **infographics** is too generic to describe these drawings because they have strong geographic components. The quality and detail of these maps make them fantastic examples of geographic information analysis, even by today's standards. Minard released his masterpiece in 1869:

> *"La carte figurative des pertes successives en hommes de l'Armeé Française dans la campagne de Russie 1812-1813," which is translated as "Figurative map of the successive losses of men of the French army in the Russian Campaign 1812-13."*

This depicts the decimation of Napoleon's army in the Russian campaign of 1812. The map shows the size and location of the army over time, along with prevailing weather conditions. The following graphic contains four different series of information on a single theme. It is a fantastic example of geographic analysis using pen and paper. The size of the army is represented by the widths of the brown and black swaths at a ratio of one millimeter for every 10,000 men. The numbers are also written along the swaths. The brown-colored path shows soldiers who entered Russia, while the black-colored path represents the ones who made it out. The map scale is shown to the right in the center as one French league (2.75 miles or 4.4 kilometers). The chart at the bottom runs from right to left and depicts the brutal freezing temperatures that were experienced by the soldiers on the return march home from Russia:

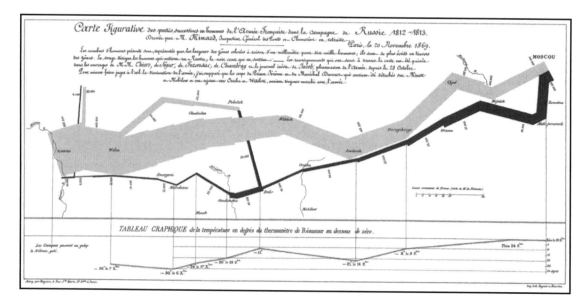

While far more mundane than a war campaign, Minard released another compelling map cataloging the number of cattle sent to Paris from around France. Minard used pie charts of varying sizes in the regions of France to show each area's variety and volume of cattle that was shipped:

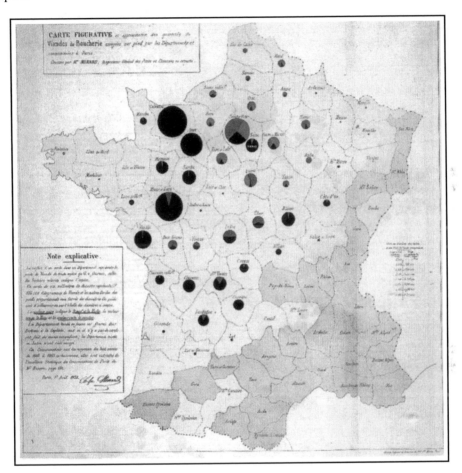

In the early 1900s, mass printing drove the development of the concept of map layers – a key feature of geospatial analysis. Cartographers drew different map elements (vegetation, roads, and elevation contours) on plates of glass that could then be stacked and photographed to be printed as a single image. If the cartographer made a mistake, only one plate of glass had to be changed instead of the entire map. Later, the development of plastic sheets made it even easier to create, edit, and store maps in this manner. However, the layering concept for maps as a benefit to analysis would not come into play until the modern computer age.

GIS

Computer mapping evolved with the computer itself in the 1960s. However, the origin of the term GIS began with the Canadian Department of Forestry and Rural Development. Dr. Roger Tomlinson headed a team of 40 developers in an agreement with IBM to build the **Canada Geographic Information System (CGIS)**. The CGIS tracked the natural resources of Canada and allowed the profiling of those features for further analysis. The CGIS stored each type of land cover as a different layer.

It also stored data in a Canadian-specific coordinate system, suitable for the entire country, which was devised for optimal area calculations. While the technology that was used was primitive by today's standards, the system had phenomenal capability at that time. The CGIS included software features that seem quite modern:

- Map projection switching
- The rubber sheeting of scanned images
- Map scale change
- Line smoothing and generalization to reduce the number of points in a feature
- Automatic gap closing for polygons
- Area measurement
- The dissolving and merging of polygons
- Geometric buffering
- The creation of new polygons
- Scanning
- The digitizing of new features from the reference data

The National Film Board of Canada produced a documentary in 1967 on the CGIS, which can be viewed at the following URL: `https://youtu.be/ 3VLGvWEuZxI`.

Tomlinson is often called the father of GIS. After launching the CGIS, he earned his doctorate from the University of London with his 1974 dissertation, entitled *The application of electronic computing methods and techniques to the storage, compilation, and assessment of mapped data*, which describes GIS and geospatial analysis. Tomlinson now runs his own global consulting firm, Tomlinson Associates Ltd., and he remains an active participant in the industry. He is often found delivering keynote addresses at geospatial conferences.

CGIS is the starting point of geospatial analysis, as defined by this book. However, this book would not have been written if not for the work of Howard Fisher and the Harvard Laboratory for Computer Graphics and Spatial Analysis at the Harvard Graduate School of Design. His work on the SYMAP GIS software, which outputs maps to a line printer, started an era of development at the laboratory, which produced two other important packages and, as a whole, permanently defined the geospatial industry. SYMAP led to other packages, including GRID and the Odyssey project, which come from the same lab:

- GRID was a raster-based GIS system that used cells to represent geographic features instead of geometry. GRID was written by Carl Steinitz and David Sinton. The system later became IMGRID.
- Odyssey was a team effort led by Nick Chrisman and Denis White. It was a system of programs that included many advanced geospatial data management features that are typical of modern geodatabase systems. Harvard attempted to commercialize these packages with limited success. However, their impact is still seen today.

Virtually, every existing commercial and open source package owes something to these code bases.

Howard Fisher produced a 1967 film using the output from SYMAP to show the urban expansion of Lansing, Michigan, from 1850 to 1965 by hand-coding decades of property information into the system. This analysis took months, but now it would take only a few minutes to recreate them because of modern tools and data.

 You can watch the film at `https://www.youtube.com/watch?v=xj8DQ7IQ8_o`.

There are now dozens of **graphical user interface (GUI)** geospatial desktop applications available today from companies including Esri, ERDAS, Intergraph, ENVI, and so on. Esri is the oldest, continuously operating GIS software company, which started in the late 1960s. In the open source realm, packages including **Quantum GIS (QGIS)** and the **Geographic Resources Analysis Support System (GRASS)** are widely used. Beyond comprehensive desktop software packages, software libraries for the building of new software exist in the thousands.

GIS can provide detailed information about the Earth, but it is still just a model. Sometimes, we need a direct representation in order to gain knowledge about current or recent changes on our planet. At that point, we need remote sensing.

Remote sensing

Remote sensing is where you collect information about an object without making physical contact with that object. In the context of geospatial analysis, that object is usually the Earth. Remote sensing also includes processing the collected information. The potential of geographic information systems is limited only by the available geographic data. The cost of land surveying, even using a modern GPS to populate a GIS, has always been resource-intensive.

The advent of remote sensing not only dramatically reduced the cost of geospatial analysis but took the field in entirely new directions. In addition to powerful reference data for GIS systems, remote sensing has made the automated and semi-automated generation of GIS data possible by extracting features from images and geographic data. The eccentric French photographer, Gaspard-Félix Tournachon, also known as Nadar, took the first aerial photograph in 1858, from a hot air balloon over Paris:

The value of a true bird's-eye view of the world was immediately apparent. As early as 1920, books on aerial photo interpretation began to appear.

When the United States entered the Cold War with the Soviet Union after World War II, aerial photography to monitor military capability became prolific with the invention of the American U-2 spy plane. The U-2 spy plane could fly at 75,000 feet, putting it out of the range of existing anti-aircraft weapons designed to reach only 50,000 feet. The American U-2 flights over Russia ended when the Soviets finally shot down a U-2 and captured the pilot.

However, aerial photography had little impact on modern geospatial analysis. Planes could only capture small footprints of an area. Photographs were tacked to walls or examined on light tables but not in the context of other information. Though extremely useful, aerial photo interpretation was simply another visual perspective.

The game changer came on October 4, 1957, when the Soviet Union launched the Sputnik 1 satellite. The Soviets had scrapped a much more complex and sophisticated satellite prototype because of manufacturing difficulties. Once corrected, this prototype would later become Sputnik 3. Instead, they opted for a simple metal sphere with four antennae and a simple radio transmitter. Other countries, including the United States, were also working on satellites. These satellite initiatives were not entirely a secret. They were driven by scientific motives as part of the **International Geophysical Year (IGY)**.

Advancement in rocket technology made artificial satellites a natural evolution for Earth science. However, in nearly every case, each country's defense agency was also heavily involved. Similar to the Soviets, other countries were struggling with complex satellite designs packed with scientific instruments. The Soviets' decision to switch to the simplest possible device was for the sole reason of launching a satellite before the Americans were effective. Sputnik was visible in the sky as it passed over, and its radio pulse could be heard by amateur radio operators. Despite Sputnik's simplicity, it provided valuable scientific information that could be derived from its orbital mechanics and radiofrequency physics.

The Sputnik program's biggest impact was on the American space program. America's chief adversary had gained a tremendous advantage in the race to space. The United States ultimately responded with the Apollo moon landings. However, before this, the US launched a program that would remain a national secret until 1995. The classified CORONA program resulted in the first pictures from space. The US and Soviet Union had signed an agreement to end spy plane flights, but satellites were conspicuously absent from the negotiations.

The following map shows the CORONA process. The dashed lines are the satellite flight paths, the long white tubes are the satellites, the smaller white cones are the film canisters, and the black blobs are the control stations that triggered the ejection of the film so that a plane could catch it in the sky:

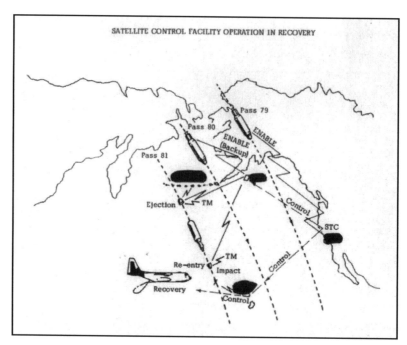

The first CORONA satellite was a 4-year effort with many setbacks. However, the program ultimately succeeded. The difficulty with satellite imaging, even today, is retrieving the images from space. The CORONA satellites used canisters of black and white film that were ejected from the vehicle once exposed. As a film canister parachuted to Earth, a US military plane would catch the package in midair. If the plane missed the canister, it would float for a brief period of time in the water before sinking into the ocean to protect the sensitive information.

The US continued to develop the CORONA satellites until they matched the resolution and photographic quality of the U-2 spy plane photos. The primary disadvantages of the CORONA instruments were reusability and timeliness. Once out of film, a satellite could no longer be of service. Additionally, the film's recovery was on a set schedule, making the system unsuitable for monitoring real-time situations. The overall success of the CORONA program, however, paved the way for the next wave of satellites, which ushered in the modern era of remote sensing.

Due to the CORONA program's secret status, its impact on remote sensing was indirect. Photographs of the Earth taken on manned US space missions inspired the idea of a civilian-operated remote sensing satellite. The benefits of such a satellite were clear, but the idea was still controversial. Government officials questioned whether a satellite was as cost-efficient as aerial photography. The military was worried that the public satellite could endanger the secrecy of the CORONA program. Other officials worried about the political consequences of imaging other countries without permission. However, the **Department of the Interior (DOI)** finally won permission for NASA to create a satellite to monitor Earth's surface resources.

On July 23, 1972, NASA launched the **Earth Resources Technology Satellite (ERTS)**. The ERTS was quickly renamed **Landsat 1**. The platform contained two sensors. The first was the **Return Beam Vidicon (RBV)** sensor, which was essentially a video camera. It was built by the radio and television giant known as the **Radio Corporation of America (RCA)**. The RBV immediately had problems, which included disabling the satellite's altitude guidance system. The second attempt at a satellite was the highly experimental **Multispectral Scanner (MSS)**. The MSS performed flawlessly and produced superior results than the RBV. The MSS captured four separate images at four different wavelengths of the light reflected from the Earth's surface.

This sensor had several revolutionary capabilities. The first and most important capability was the first global imaging of the planet scanning every spot on the Earth every 16 days. The following image from NASA illustrates this flight and collection pattern, which is a series of overlapping swaths as the sensor orbits the Earth, capturing tiles of data each time the sensor images a location on the Earth:

It also recorded light beyond the visible spectrum. While it did capture green and red light visible to the human eye, it also scanned near-infrared light at two different wavelengths not visible to the human eye. The images were stored and transmitted digitally to three different ground stations in Maryland, California, and Alaska. Its multispectral capability and digital format meant that the aerial view provided by Landsat wasn't just another photograph from the sky. It was beaming down the data. This data could be processed by computers to output derivative information about the Earth in the same way a GIS provided derivative information about the Earth by analyzing one geographic feature in the context of another. NASA promoted the use of Landsat worldwide and made the data available at very affordable prices to anyone who asked.

This global imaging capability led to many scientific breakthroughs, including the discovery of previously unknown geography, which occurred as late as 1976. For example, using Landsat imagery, the Government of Canada located a tiny uncharted island inhabited by polar bears. They named the new landmass Landsat Island.

Landsat 1 was followed by six other missions that were turned over to the **National Oceanic and Atmospheric Administration (NOAA)** as the responsible agency. Landsat 6 failed to achieve orbit due to a ruptured manifold, which disabled its maneuvering engines. During some of these missions, the satellites were managed by the **Earth Observation Satellite (EOSAT)** company, now called **Space Imaging**, but returned to government management by the Landsat 7 mission. The following image from NASA is a sample of a Landsat 7 product:

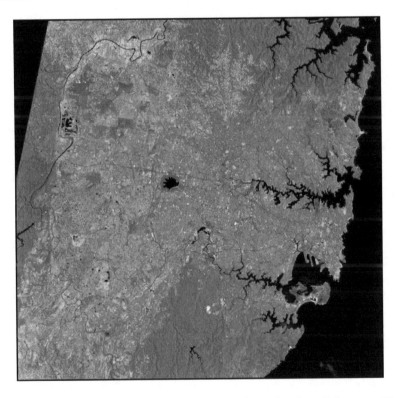

The **Landsat Data Continuity Mission (LDCM)** was launched on February 13, 2013, and began collecting images on April 27, 2013, as part of its calibration cycle to become Landsat 8. The LDCM is a joint mission between NASA and the **US Geological Survey (USGS)**.

Elevation data

Remote sensing data can measure the Earth in two dimensions. But we can also use remote sensing to measure the Earth in three dimensions using digital elevation data, which we include in a Digital Elevation Model. A **Digital Elevation Model (DEM)** is a three-dimensional representation of a planet's terrain. In the context of this book, this planet is the Earth. The history of digital elevation models is far less complicated than remotely-sensed imagery but no less significant. Before computers, representations of elevation data were limited to topographic maps created through traditional land surveys. The technology existed to create three-dimensional models from stereoscopic images or physical models from materials such as clay or wood, but these approaches were not widely used for geography.

The concept of digital elevation models came about in 1986 when the French space agency, **Centre National d'études Spatiales (CNES)** or **National Center for the Study of Space**, launched its SPOT-1 satellite, which included a stereoscopic radar. This system created the first usable DEM. Several other US and European satellites followed this model with similar missions.

In February 2000, Space Shuttle Endeavour conducted the **Shuttle Radar Topography Mission (SRTM)**, which collected elevation data of over 80% of the Earth's surface using a special radar antenna configuration that allowed a single pass. This model was surpassed in 2009 by the joint US and Japanese mission using the **Advanced Spaceborne Thermal Emission and Reflection Radiometer (ASTER)** sensor aboard NASA's Terra satellite. This system captured 99% of the Earth's surface but has proven to have minor data issues. Since the Space Shuttle's orbit did not cross the Earth's poles, it did not capture the entire surface. SRTM remains the gold standard. The following image from the USGS (`https://www.usgs.gov/media/images/national-elevation-dataset`) shows a colorized DEM known as a hillshade. Greener areas are lower elevations, while yellow and brown areas are mid-range to high elevations:

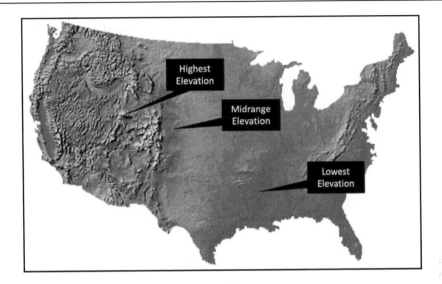

Recently, more ambitious attempts at a worldwide elevation dataset are underway in the form of the TerraSAR-X and TanDEM-X satellites, which were launched by Germany in 2007 and 2010, respectively. These two radar elevation satellites worked together to produce a global DEM called WorldDEM that was released on April 15, 2014. This dataset has a relative accuracy of 2 meters and an absolute accuracy of 4 meters.

Computer-aided drafting

Computer-aided drafting (CAD) is worth mentioning, though it does not directly relate to geospatial analysis. The history of CAD system development parallels and intertwines with the history of geospatial analysis. CAD is an engineering tool used to model two- and three-dimensional objects, usually for engineering and manufacturing. The primary difference between a geospatial model and CAD model is that a geospatial model is referenced to the Earth, whereas a CAD model can possibly exist in abstract space.

For example, a three-dimensional blueprint of a building in a CAD system would not have latitude or longitude, but in a GIS, the same building model would have a location on the Earth. However, over the years, CAD systems have taken on many features of GIS systems and are commonly used for smaller GIS projects. Likewise, many GIS programs can import CAD data that has been georeferenced. Traditionally, CAD tools were designed primarily to engineer data that was not geospatial.

However, engineers who became involved with geospatial engineering projects, such as designing a city's utility electric system, would use the CAD tools that they were familiar with in order to create maps. Over time, the GIS software evolved to import the geospatial-oriented CAD data produced by engineers, and a CAD tools evolved to support geospatial data creation and better compatibility with GIS software. AutoCAD by Autodesk and ArcGIS by Esri were the leading commercial packages to develop this capability, and the **Geospatial Data Abstraction Library (GDAL)** OGR library developers added CAD support as well.

Geospatial analysis and computer programming

Modern geospatial analysis can be conducted with the click of a button in any of the easy-to-use commercial or open source geospatial packages. So, why would you want to use a programming language to learn this field? The most important reasons are as follows:

- You want complete control of the underlying algorithms, data, and execution
- You want to automate specific, repetitive analysis tasks with minimal overhead from a large, multipurpose geospatial framework
- You want to create a program that's easy to share
- You want to learn geospatial analysis beyond pushing buttons in software

The geospatial industry is gradually moving away from the traditional workflow, in which teams of analysts use expensive desktop software to produce geospatial products. Geospatial analysis is being pushed toward automated processes that reside in the cloud. End-user software is moving toward task-specific tools, many of which are accessed from mobile devices. Knowledge of geospatial concepts and data, as well as the ability to build custom geospatial processes, is where the geospatial work in the near future lies.

Object-oriented programming for geospatial analysis

Object-oriented programming is a software development paradigm in which concepts are modeled as objects that have properties and behaviors represented as attributes and methods, respectively. The goal of this paradigm is more modular software in which one object can inherit from one or more other objects to encourage software reuse.

The Python programming language is known for its ability to serve multiple roles as a well-designed, object-oriented language, a procedural scripting language, or even a functional programming language. However, you never completely abandon object-oriented programming in Python because even its native data types are objects and all Python libraries, known as modules, adhere to a basic object structure and behavior.

Geospatial analysis is the perfect activity for object-oriented programming. In most object-oriented programming projects, the objects are abstract concepts, such as database connections that have no real-world analogy. However, in geospatial analysis, the concepts that are modeled are, well, real-world objects! The domain of geospatial analysis is the Earth and everything on it. Trees, buildings, rivers, and people are all examples of objects within a geospatial system.

A common example in literature for newcomers to object-oriented programming is the concrete analogy of a cat. Books on object-oriented programming frequently use some form of the following example.

Imagine that you are looking at a cat. We know some information about the cat, such as its name, age, color, and size. These features are the properties of the cat. The cat also exhibits behaviors such as eating, sleeping, jumping, and purring. In object-oriented programming, objects have properties and behaviors too. You can model a real-world object such as the cat in our example, or something more abstract such as a bank account.

Most concepts in object-oriented programming are far more abstract than the simple cat paradigm or even a bank account. However, in geospatial analysis, the objects that are modeled remain concrete, such as the simple cat analogy, and in many cases are cats. Geospatial analysis allows you to continue with the simple cat analogy and even visualize it. The following map represents the feral cat population of Australia using data provided by the **Atlas of Living Australia (ALA)**:

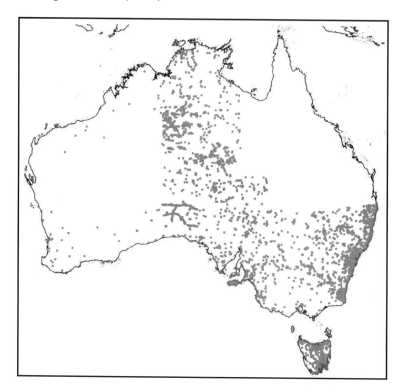

So, we can use computers to analyze the relationships between of features on Earth, but why should we? In the next section, we'll look at why geospatial analysis is a worthwhile endeavor.

The importance of geospatial analysis

Geospatial analysis helps people make better decisions. It doesn't make the decision for you, but it can answer critical questions that are at the heart of the choice to be made and often cannot be answered any other way. Until recently, geospatial technology and data were tools available only to governments and well-funded researchers. However, in the last decade, data has become much more widely available and software has become much more accessible to anyone.

In addition to freely available government satellite imagery, many local governments now conduct aerial photo surveys and make the data available online. The ubiquitous Google Earth provides a cross-platform spinning globe view of the Earth with satellite and aerial data, streets, points of interest, photographs, and much more. Google Earth users can create custom **Keyhole Markup Language** (**KML**) files, which are XML files that are used to load and style data to the globe. This program and similar tools are often called geographic exploration tools because they are excellent data viewers but provide very limited data analysis capabilities.

The ambitious OpenStreetMap project (`https://www.openstreetmap.org/#map=5/51.500/-0.100`) is a crowd-sourced, worldwide, geographic base map containing most layers commonly found in a GIS. Nearly every mobile phone now contains a GPS, along with mobile apps to collect GPS tracks as points, lines, or polygons. Most phones will also tag photos taken with the phone's camera with GPS coordinates. In short, anyone can be a geospatial analyst.

The global population has reached 7 billion people. The world is changing faster than ever before. The planet is undergoing environmental changes that have never been seen in recorded history. Faster communication and transportation increase the interaction between us and the environment in which we live. Managing people and resources safely and responsibly is more challenging than ever. Geospatial analysis is the best approach to understanding our world more efficiently and deeply. The more politicians, activists, relief workers, parents, teachers, first responders, medical professionals, and small businesses that harness the power of geospatial analysis, the more potential we have for a better, healthier, safer, and fairer world.

GIS concepts

In order to begin geospatial analysis, we need to understand some key underlying concepts that are unique to the field. The list isn't long, but nearly every aspect of analysis traces back to one of these ideas.

Thematic maps

As its name suggests, a thematic map portrays a specific theme. A general reference map visually represents features as they relate geographically to navigation or planning. A thematic map goes beyond location to provide the geographic context for information around a central idea. Usually, a thematic map is designed for a targeted audience to answer specific questions. The value of thematic maps lies in what they do not show. A thematic map will use minimal geographic features to avoid distracting the reader from the theme. Most thematic maps include political boundaries such as country or state borders but omit navigational features, such as street names or points of interest beyond major landmarks that orient the reader.

The cholera map by Dr. John Snow earlier in this chapter is a perfect example of a thematic map. Common uses for thematic maps are visualizing health issues, such as disease, election results, and environmental phenomena such as rainfall. These maps are also the most common output of geospatial analysis. The following map from the United States Census Bureau shows cancer mortality rates by state:

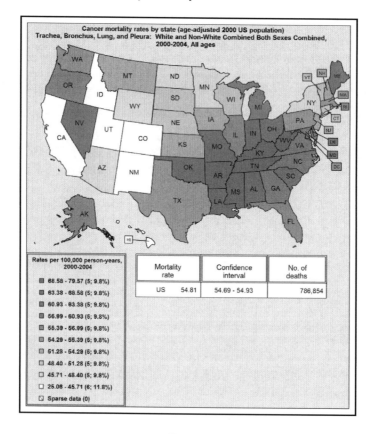

Thematic maps tell a story and are very useful. However, it is important to remember that, while thematic maps are models of reality just like any other map, they are also generalizations of information. Two different analysts using the same source of information will often come up with very different thematic maps, depending on how they analyze and summarize the data. They may also choose to focus on different aspects of the dataset. The technical nature of thematic maps often leads people to treat them as if they are scientific evidence. However, geospatial analysis is often inconclusive. While the analysis may be based on scientific data, the analyst does not always follow the rigor of the scientific method.

In his classic book, *How to Lie with Maps*, Mark Monmonier, University of Chicago Press, demonstrates in detail how maps are easily manipulated models of reality, which are commonly abused. This fact doesn't degrade the value of these tools. The legendary statistician, George Box, wrote the following in his 1987 book, *Empirical Model-Building and Response Surfaces*:

> *"Essentially, all models are wrong, but some are useful."*

Thematic maps have been used as guides to start (and end) wars, stop deadly diseases in their tracks, win elections, feed nations, fight poverty, protect endangered species, and rescue those impacted by disaster. Thematic maps may be the most useful models ever created.

In its purest form, a database is simply an organized collection of information. A **database management system (DBMS)** is an interactive suite of software that can interact with a database. People often use the word database as a catch-all term referring to both the DBMS and underlying data structure. Databases typically contain alphanumeric data and, in some cases, binary large objects or blobs, which can store binary data such as images. Most databases also allow a relational database structure in which entries in normalized tables can be referenced to each other in order to create many-to-one and one-to-many relationships among data.

Spatial databases, also known as geodatabases, use specialized software to extend a traditional **relational database management system (RDBMS)** to store and query data defined in a two-dimensional or three-dimensional space. Some systems also account for a series of data over time. In a spatial database, attributes about geographic features are stored and queried as traditional relational database structures. These spatial extensions allow you to query geometries using **Structured Query Language (SQL)** in a similar way to traditional database queries. Spatial queries and attribute queries can also be combined to select results based on both location and attributes.

Spatial indexing

Spatial indexing is a process that organizes geospatial vector data for faster retrieval. It is a way of prefiltering the data for common queries or rendering. Indexing is commonly used in large databases to speed up the returns to queries. Spatial data is no different. Even a moderately sized geodatabase can contain millions of points or objects. If you perform a spatial query, every point in the database must be considered by the system in order to include it or eliminate it in the results. Spatial indexing groups data in ways that allow large portions of the dataset to be eliminated from consideration by doing computationally simpler checks before going into a detailed and slower analysis of the remaining items.

Metadata

Metadata is defined as data about data. Accordingly, geospatial metadata is data about geospatial datasets that provides traceability for the source and history of a dataset, as well as a summary of the technical details. Metadata also provides long-term preservation of data by way of documenting the asset over time.

Geospatial metadata can be represented by several possible standards. One of the most prominent standards is the international standard, **ISO 19115-1**, which includes hundreds of potential fields to describe a single geospatial dataset. Additionally, the **ISO 19115-2** standard includes extensions for geospatial imagery and gridded data. Some example fields include spatial representation, temporal extent, and lineage. ISO 19115-3 is the standard for describing the procedure to generate an XML schema from ISO geographic metadata. Dublin Core is another international standard that was developed for digital data that has been extended for geospatial data, along with the associated DCAT vocabulary for building catalogs of data from a single source.

The primary use of metadata is for cataloging datasets. Modern metadata can be ingested by geographic search engines, making it potentially discoverable by other systems automatically. It also lists points of contact for a dataset if you have questions. Metadata is an important support tool for geospatial analysts and adds credibility and accessibility to your work. The **Open Geospatial Consortium** (**OGC**), which created the **Catalog Service for the Web** (**CSW**), is used to manage metadata. The `pycsw` Python library implements the CSW standard.

 Metadata is an important documentation tool that's used to manage geospatial data, while `pycsw` is an OGC-compliant CSW implementation. You can learn more about `pycsw` at `https://pycsw.org`.

Map projections

Map projections have entire books devoted to them and can be a challenge for new analysts. If you take any 3D object and flatten it on a plane, such as your screen or a sheet of paper, the object is distorted. Many grade school geography classes demonstrate this concept by having students peel an orange and then attempt to lay the peel flat on their desk in order to understand the resulting distortion. The same effect occurs when you take the round shape of the Earth and project it onto a computer screen.

In geospatial analysis, you can manipulate this distortion to preserve common properties, such as area, scale, bearing, distance, or shape. There is no one-size-fits-all solution to map projections. The choice of projection is always a compromise of gaining accuracy in one dimension in exchange for error in another. Projections are typically represented as a set of over 40 parameters, as either XML or in a text format called **Well-Known Text (WKT)**, which is used to define the transformation algorithm.

The **International Association of Oil and Gas Producers (IOGP)** maintains a registry of the most well-known projections. The organization was formerly known as the **European Petroleum Survey Group (EPSG)**. The entries in the registry are still known as EPSG codes. The EPSG maintained the registry as a common benefit for the oil and gas industry, which is a prolific user of geospatial analysis for energy exploration. At the last count, this registry contained over 5,000 entries.

As recently as 10 years ago, map projections were of primary concern for a geospatial analyst. Data storage was expensive, high-speed internet was rare, and cloud computing didn't really exist. Geospatial data was typically exchanged among small groups working in separate areas of interest. The technology constraints at the time meant that geospatial analysis was highly localized. Analysts would use the best projection for their area of interest.

Data in different projections could not be displayed on the same map because they represent two different models of the Earth. Any time an analyst received data from a third party, it had to be reprojected before they could use it with the existing data. This process was tedious and time-consuming.

Most geospatial data formats do not provide a way to store the projection information. This information is stored in an ancillary file, usually as text or XML. Since analysts didn't exchange data often, many people wouldn't bother defining projection information. Every analyst's nightmare was to come across an extremely valuable dataset that was missing the projection information. It rendered the dataset useless. The coordinates in the file are just numbers and offer no clue about the projection. With over 5,000 choices, it was nearly impossible to guess.

Now, thanks to modern software and the internet making data exchange easier and more common, nearly every data format has added a metadata format that defines a projection or places it in the file header, if supported. Advances in technology have also allowed for global base maps, which allow for more common uses of projections, such as the common Google Mercator projection that's used for Google Maps. This projection is also known as Web Mercator and uses code EPSG:3857 (or the deprecated EPSG:900913).

Geospatial portal projects such as OpenStreetMap (`https://www.openstreetmap.org/#map=5/51.500/-0.100`) and NationalAtlas.gov have consolidated datasets for much of the world in common projections. Modern geospatial software can also reproject data on the fly, saving the analyst the trouble of preprocessing the data before using it. Closely related to map projections are geodetic datums. A **datum** is a model of the Earth's surface that's used to match the location of features on the Earth to a coordinate system. One common datum is called WGS 84 and is used by GPS.

Rendering

The exciting part of geospatial analysis is visualization. Since geospatial analysis is a computer-based process, it is good to be aware of how geographic data appears on a computer screen.

Geographic data including points, lines, and polygons are stored numerically as one or more points, which come in (x,y) pairs or (x,y,z) tuples. The x represents the horizontal axis on a graph, while the y represents the vertical axis. The z represents terrain elevation. In computer graphics, a computer screen is represented by an x- and y-axis. The z-axis is not used because the computer screen is treated as a two-dimensional plane by most graphics software APIs. However, because desktop computing power continues to improve, three-dimensional maps are starting to become more common.

Another important factor is screen coordinates versus world coordinates. Geographic data is stored in a coordinate system representing a grid overlaid on the Earth, which is three-dimensional and round. Screen coordinates, also known as pixel coordinates, represent a grid of pixels on a flat, two-dimensional computer screen. Mapping x and y world coordinates to pixel coordinates is fairly straightforward and involves a simple scaling algorithm. However, if a z coordinate exists, then a more complicated transformation must be performed to map coordinates from a three-dimensional space to a two-dimensional plane. These transformations can be computationally costly and therefore slow if not handled correctly.

In the case of remote sensing data, the challenge is typically the file size. Even a moderately sized satellite image that is compressed can be tens, if not hundreds, of megabytes. Images can be compressed using two methods:

- **Lossless methods**: They use tricks to reduce the file size without discarding any data
- **Lossy compression algorithms**: They reduce the file size by reducing the amount of data in the image while avoiding a significant change in the appearance of the image

Rendering an image on the screen can be computationally intensive. Most remote sensing file formats allow for the storing of multiple lower-resolution versions of the image – called **overviews** or **pyramids** – for the sole purpose of faster rendering at different scales. When zoomed out from the image to a scale where you can see the detail of the full resolution image, a preprocessed, lower-resolution version of the image is displayed quickly and seamlessly.

Remote sensing concepts

Most of the GIS concepts we've described also apply to raster data. However, raster data has some unique properties as well. Earlier in this chapter, when we went over the history of remote sensing, the focus was on Earth imaging from aerial platforms. It is important to note that raster data can come in many forms, including ground-based radar, laser range finders, and other specialized devices to detect gases, radiation, and other forms of energy in a geographic context.

For the purpose of this book, we will focus on remote sensing platforms that capture large amounts of Earth data. These sources included Earth imaging systems, certain types of elevation data, and some weather systems, where applicable.

Images as data

Raster data is captured digitally as square tiles. This means that the data is stored on a computer as a numerical array of rows and columns. If the data is multispectral, the dataset will usually contain multiple arrays of the same size, which are geospatially referenced together to represent a single area on the Earth. These different arrays are called **bands**. Any numerical array can be represented on a computer as an image. In fact, all computer data is ultimately numbers. In geospatial analysis, it is important to think of images as a numeric array because mathematical formulas are used to process them.

In remotely sensed images, each pixel represents both space (the location on the Earth of a certain size) and the reflectance captured as light reflected from the Earth at that location into space. So, each pixel has a ground size and contains a number representing the intensity. Since each pixel is a number, we can perform mathematical equations on this data to combine data from different bands and highlight specific classes of objects in the image. If the wavelength value is beyond the visible spectrum, we can highlight features that aren't visible to the human eye. Substances such as chlorophyll in plants can be greatly contrasted using a specific formula called the **Normalized Difference Vegetation Index (NDVI)**.

By processing remotely sensed images, we can turn this data into visual information. Using the NDVI formula, we can answer the question, *what is the relative health of the plants in this image?* You can also create new types of digital information, which can be used as input for computer programs to output other types of information.

Remote sensing and color

Computer screens display images as combinations of **Red, Green, and Blue (RGB)** to match the capability of the human eye. Satellites and other remote sensing imaging devices can capture light beyond this visible spectrum. On a computer, wavelengths beyond the visible spectrum are represented in the visible spectrum so that we can see them. These images are known as **false color** images. In remote sensing, for instance, infrared light makes moisture highly visible.

This phenomenon has a variety of uses, such as monitoring ground saturation during a flood or finding hidden leaks in a roof or levee.

Common vector GIS concepts

In this section, we will discuss the different types of GIS processes that are commonly used in geospatial analysis. This list is not exhaustive; however, it provides you with the essential operations that all other operations are based on. If you understand these operations, you will quickly understand much more complex processes as they are either derivatives or combinations of these processes.

Data structures

GIS vector data uses coordinates consisting of, at a minimum, an x horizontal value and a y vertical value to represent a location on the Earth. In many cases, a point may also contain a z value. Other ancillary values are possible, including measurements or timestamps.

These coordinates are used to form points, lines, and polygons to model real-world objects. Points can be a geometric feature in and of themselves or they can connect line segments. Closed areas created by line segments are considered polygons. Polygons model objects such as buildings, terrain, or political boundaries.

A GIS feature can consist of a single point, line, or polygon, or it can consist of more than one shape. For example, in a GIS polygon dataset containing world country boundaries, the Philippines, which is made up of 7,107 islands, would be represented as a single country made up of thousands of polygons.

Vector data typically represents topographic features better than raster data. Vector data has more accuracy potential and is more precise. However, collecting vector data on a large scale is also traditionally more costly than raster data.

Two other important terms related to vector data structures are bounding box and convex hull. The bounding box, or minimum bounding box, is the smallest possible square that contains all of the points in a dataset. The following diagram demonstrates a bounding box for a collection of points:

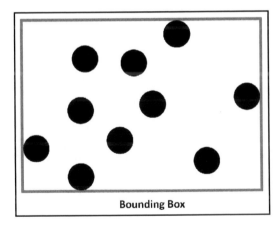

Bounding Box

The convex hull of a dataset is similar to the bounding box, but instead of a square, it is the smallest possible polygon that can contain a dataset. The following diagram shows the same point data as the previous example, with the convex hull polygon shown in red:

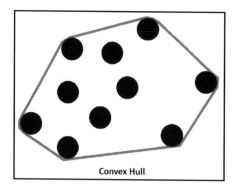

Convex Hull

As you can see, the bounding box of a dataset always contains a convex hull.

Geospatial rules about polygons

In geospatial analysis, there are several general rules of thumb regarding polygons that are different from mathematical descriptions of polygons:

- Polygons must have at least four points – the first and last points must be the same
- A polygon boundary should not overlap itself
- Polygons in a layer shouldn't overlap
- A polygon in a layer inside another polygon is considered a hole in the underlying polygon

Different geospatial software packages and libraries handle exceptions to these rules differently, which can lead to confusing errors or software behaviors. The safest route is to make sure that your polygons obey these rules. There's one more important piece of information about polygons that we need to talk about.

A polygon is, by definition, a closed shape, which means that the first and last vertices of a polygon are identical. Some geospatial software will throw an error if you haven't explicitly duplicated the first point as the last point in the polygon dataset. Other software will automatically close the polygon without complaining. The data format that you use to store your geospatial data may also dictate how polygons are defined. This issue is a gray area and so it didn't make the polygon rules, but knowing this quirk will come in handy someday when you run into an error that you can't explain easily.

Buffer

A buffer operation can be applied to spatial objects, including points, lines, or polygons. This operation creates a polygon around the object at a specified distance. Buffer operations are used for proximity analysis: for example, establishing a safety zone around a dangerous area. Let's review this diagram:

The black shapes represent the original geometry, while the red outlines represent the larger buffer polygons that were generated from the original shape.

Dissolve

A dissolve operation creates a single polygon out of adjacent polygons. Dissolves are also used to simplify data that's been extracted from remote sensing, as shown here:

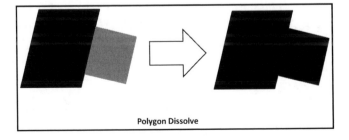

A common use for a dissolve operation is to merge two adjacent properties in a tax database that has been purchased by a single owner.

Generalize

Objects that have more points than necessary for the geospatial model can be generalized to reduce the number of points that are used to represent the shape. This operation usually requires a few attempts to get the optimal number of points without compromising the overall shape. It is a data optimization technique that's used to simplify data for the efficiency of computing or better visualization. This technique is useful in web mapping applications.

Here is an example of a polygon generalization:

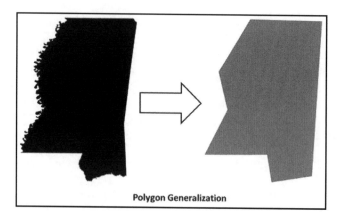

Polygon Generalization

Since computer screens have a resolution of 72 dots per inch (dpi), highly detailed point data, which would not be visible, can be reduced so that less bandwidth is used to send a visually equivalent map to the user.

Intersection

An intersection operation is used to see if one part of a feature intersects with one or more features. This operation is used for spatial queries in proximity analysis and is often a follow-on operation to buffer analysis:

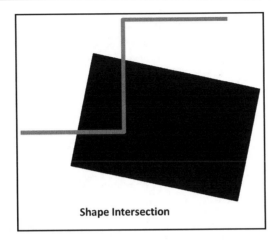

Shape Intersection

A merge operation combines two or more non-overlapping shapes in a single multi-shape object. Multi-shape objects are shapes that maintain separate geometries but are treated as a single feature with a single set of attributes by the GIS:

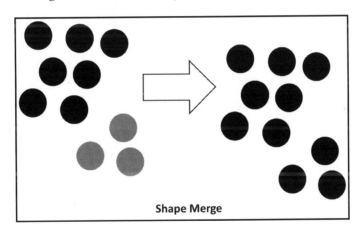

Shape Merge

A fundamental geospatial operation is checking to see whether a point is inside a polygon. This one operation is the atomic building block of many different types of spatial queries. If the point is on the boundary of the polygon, it is considered inside. Very few spatial queries exist that do not rely on this calculation in some way. However, it can be very slow on a large number of points.

The most common and efficient algorithm to detect whether a point is inside a polygon is called the **ray casting algorithm**. First, a test is performed to see if the point is on the polygon boundary. Next, the algorithm draws a line from the point in question in a single direction. The program counts the number of times the line crosses the polygon boundary until it reaches the bounding box of the polygon, as shown here:

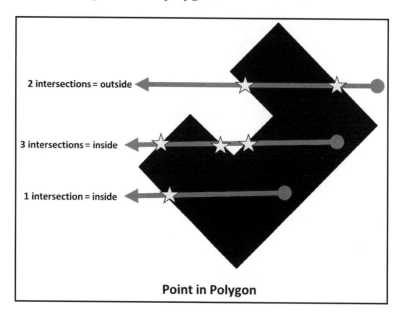

Point in Polygon

The bounding box is the smallest box that can be drawn around the entire polygon. If the number is odd, the point is inside. If the number of boundary intersections is even, the point is outside.

Union

The union operation is less common, but very useful when you wish to combine two or more overlapping polygons in a single shape. It is similar to dissolve, but in this case, the polygons are overlapping as opposed to being adjacent:

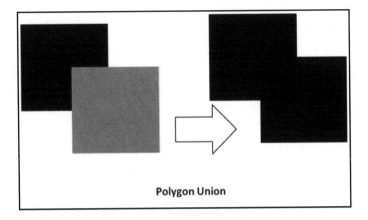

Polygon Union

Usually, this operation is used to clean up automatically generated feature datasets from remote sensing operations.

Join

A join or SQL join is a database operation that's used to combine two or more tables of information. Relational databases are designed to avoid storing redundant information for one-to-many relationships. For example, a US state may contain many cities. Rather than creating a table for each state containing all of its cities, a table of states with numeric IDs is created, while a table for all the cities in every state is created with a state numeric ID.

In a GIS, you can also have spatial joins that are part of the spatial extension software for a database. In spatial joins, you combine the attributes in the same way that you do in a SQL join. However, the relation is based on the spatial proximity of the two features.

To follow the previous cities example, we could add the county name that each city resides in using a spatial join. The cities layer could be loaded over a county polygon layer whose attributes contain the county name. The spatial join would determine which city is in which county and perform a SQL join to add the county name to each city's attribute row.

Common raster data concepts

As we mentioned earlier, remotely sensed raster data is a matrix of numbers. Remote sensing contains thousands of operations that can be performed on data. This field changes on almost a daily basis as new satellites are put into space and computer power increases.

Despite its decade-long history, we haven't even scratched the surface of the knowledge that this field can provide to the human race. Once again, similar to the common GIS processes, this minimal list of operations allows you to evaluate any technique that's used in remote sensing.

Band math

Band math is multidimensional array mathematics. In array math, arrays are treated as single units, which are added, subtracted, multiplied, and divided. However, in an array, the corresponding numbers in each row and column across multiple arrays are computed simultaneously. These arrays are termed matrices, and computations involving matrices are the focus of linear algebra.

Change detection

Change detection is the process of taking two images of the same location at different times and highlighting those changes. A change could be due to the addition of something on the ground, such as a new building, or the loss of a feature, such as coastal erosion. There are many algorithms that detect changes among images and also determine qualitative factors such as how long ago the change took place.

The following image from a research project by the US **Oak Ridge National Laboratory (ORNL)** shows rainforest deforestation between 1984 and 2000 in the state of Rondonia, Brazil:

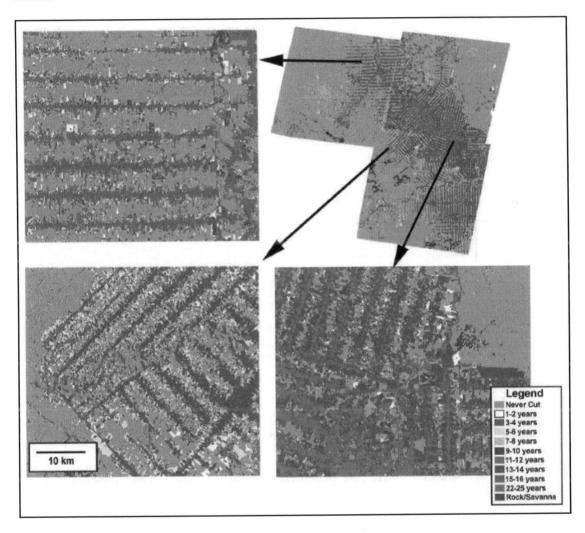

Colors are used to show how recently the forest was cut. Green represents virgin rainforests, white represents a forest that was cut within two years of the end of the date range, red represents within 22 years, and the other colors fall in-between, as described in the legend.

Histogram

A histogram is the statistical distribution of values in a dataset. The horizontal axis represents a unique value in a dataset, while the vertical axis represents the frequency of this unique value in the raster. The following example from NASA shows a histogram for a satellite image that has been classified into different categories, representing the underlying surface features:

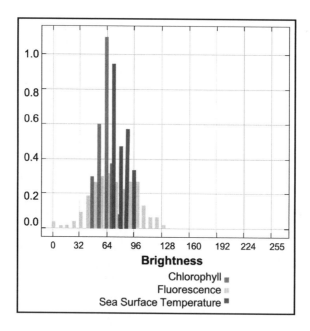

A histogram is a key operation in most raster processing. It can be used for everything from enhancing contrast in an image to serving as a basis for object classification and image comparison.

Feature extraction

Feature extraction is the manual or automatic digitization of features in an image to points, lines, or polygons. This process serves as the basis for the vectorization of images in which a raster is converted into a vector dataset. An example of feature extraction is extracting a coastline from a satellite image and saving it as a vector dataset.

If this extraction is performed over several years, you could monitor the erosion or other changes along this coastline.

Supervised and unsupervised classification

Objects on the Earth reflect different wavelengths of light, depending on the materials that they are made of. In remote sensing, analysts collect wavelength signatures for specific types of land cover (for example, concrete) and build a library for a specific area. A computer can then use this library to automatically locate classes in the library in a new image of the same area.

In unsupervised classification, a computer groups pixels with similar reflectance values in an image without any other reference information other than the histogram of the image.

Creating the simplest possible Python GIS

Now that we have a better understanding of geospatial analysis, the next step is to build a simple GIS known as SimpleGIS using Python. This small program will be a technically complete GIS with a geographic data model and the ability to render the data as a visual thematic map showing the population of different cities.

The data model will also be structured so that you can perform basic queries. Our SimpleGIS will contain the state of Colorado, three cities, and population counts for each city.

Most importantly, we will demonstrate the power and simplicity of Python programming by building this tiny system in pure Python. We will only use modules available in the standard Python distribution without downloading any third-party libraries.

Getting started with Python

As we stated earlier, this book assumes that you have some basic knowledge of Python. The only module that's used in the following example is the turtle module, which provides a very simple graphics engine based on the Tkinter library included with Python. If you used the installers for Windows or macOS, the Tkinter library should be included already. If you compiled Python yourself or are using a distribution from somewhere besides Python.org (https://www.python.org), then make sure that you can import the turtle module by typing in the following in the command prompt. This will run the turtle demo script:

```
python -m turtle
```

The preceding command will begin a real-time drawing program that will demonstrate the capabilities of the turtle module similar to the following screenshot:

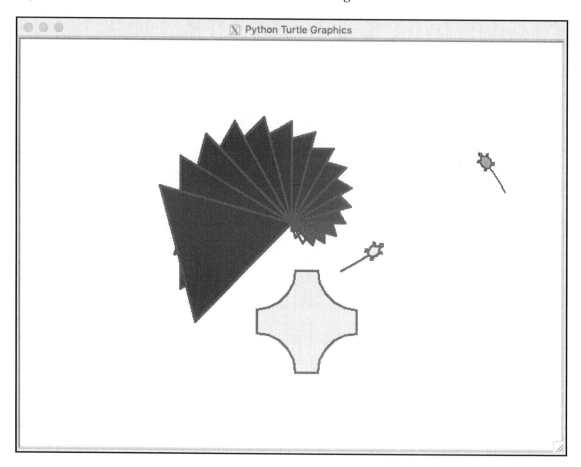

Now that we've seen what the turtle graphics module can do, let's use it to build an actual GIS!

Building a SimpleGIS

The code is divided into two different sections:

- The data model section
- The map renderer that draws the data

For the data model, we will use simple Python lists. A Python list is a native data type that serves as a container for other Python objects in a specified order. Python lists can contain other lists and are great for simple data structures. They also map well to more complex structures or even databases if you decide you want to develop your script further.

The second portion of the code will render the map using the Python turtle graphics engine. We will have only one function in the GIS that converts the world coordinates – in this case, longitude and latitude – into pixel coordinates. All graphics engines have an origin point of **(0,0)** and it's usually in the top-left or lower-left corner of the canvas.

Turtle graphics are designed to teach programming visually. The turtle graphics canvas uses an origin of **(0,0)** in the center, similar to a graphing calculator. The following graph illustrates the type of Cartesian graph that the `turtle` module uses. Some of the points are plotted in both positive and negative space:

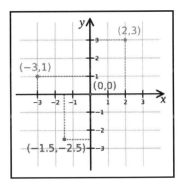

This also means that the turtle graphics engine can have negative pixel coordinates, which is uncommon for graphics canvases. However, for this example, the `turtle` module is the quickest and simplest way to render our map.

Setting up the data model

You can run this program interactively in the Python interpreter or you can save the complete program as a script and run it. The Python interpreter is an incredibly powerful way to learn about new concepts because it gives you real-time feedback on errors or unexpected program behavior. You can easily recover from these issues and try something else until you get the results that you want:

1. In Python, you usually import modules at the beginning of the script, so we'll import the `turtle` module first. We'll use Python's `import` feature to assign the module the name `t` to save space and time when typing `turtle` commands:

   ```
   import turtle as t
   ```

2. Next, we'll set up the data model, starting with some simple variables that allow us to access list indexes by name instead of numbers to make the code easier to follow. Python lists index the contained objects, starting with the number 0. So, if we wanted to access the first item in a list called `myList`, we would reference it as follows:

   ```
   myList[0]
   ```

3. To make our code easier to read, we can also use a variable name that's been assigned to commonly used indexes:

   ```
   firstItem = 0
   myList[firstItem]
   ```

   In computer science, assigning commonly used numbers to an easy-to-remember variable is a common practice. These variables are called constants. So, for our example, we'll assign constants for some common elements that are used for all of the cities. All cities will have a name, one or more points, and a population count:

   ```
   NAME = 0
   POINTS = 1
   POP = 2
   ```

4. Now, we'll set up the data for Colorado as a list with a name, polygon points, and population. Note that the coordinates are a list within a list:

   ```
   state = ["COLORADO", [[-109, 37],[-109, 41],[-102, 41],[-102, 37]],
   5187582]
   ```

5. The cities will be stored as nested lists. Each city's location consists of a single point as a longitude and latitude pair. These entries will complete our GIS data model. We'll start with an empty list called `cities` and then append the data to this list for each city:

```
cities = []
cities.append(["DENVER",[-104.98, 39.74], 634265])
cities.append(["BOULDER",[-105.27, 40.02], 98889])
cities.append(["DURANGO",[-107.88,37.28], 17069])
```

6. We will now render our GIS data as a map by first defining a map size. The width and height can be anything that you want, depending on your screen resolution:

```
map_width = 400
map_height = 300
```

7. In order to scale the map to the graphics canvas, we must first determine the bounding box of the largest layer, which is the state. We'll set the map's bounding box to a global scale and reduce it to the size of the state. To do so, we'll loop through the longitude and latitude of each point and compare it with the current minimum and maximum x and y values. If it is larger than the current maximum or smaller than the current minimum, we'll make this value the new maximum or minimum, respectively:

```
minx = 180
maxx = -180
miny = 90
maxy = -90
forx,y in state[POINTS]:
if x < minx:
    minx = x
elif x > maxx:
    maxx = x
if y < miny:
    miny = y
elif y > maxy:
    maxy = y
```

8. The second step when it comes to scaling is calculating a ratio between the actual state and the tiny canvas that we will render it on. This ratio is used for coordinate to pixel conversion. We get the size along the *x* and *y* axes of the state and then we divide the map width and height by these numbers to get our scaling ratio:

```
dist_x = maxx - minx
dist_y = maxy - miny
x_ratio = map_width / dist_x
y_ratio = map_height / dist_y
```

9. The following function, called `convert()`, is our only function in `SimpleGIS`. It transforms a point in the map coordinates from one of our data layers into pixel coordinates using the previous calculations. You'll notice that, in the end, we divide the map width and height in half and subtract it from the final conversion to account for the unusual center origin of the turtle graphics canvas. Every geospatial program has some form of this function:

```
def convert(point):
    lon = point[0]
    lat = point[1]
    x = map_width - ((maxx - lon) * x_ratio)
    y = map_height - ((maxy - lat) * y_ratio)
    # Python turtle graphics start in the
    # middle of the screen
    # so we must offset the points so they are centered
    x = x - (map_width/2)
    y = y - (map_height/2)
    return [x,y]
```

Now comes the exciting part! We're ready to render our GIS as a thematic map.

Rendering the map

The `turtle` module uses the concept of a cursor, known as a **pen**. Moving the cursor around the canvas is exactly the same as moving a pen around a piece of paper. The cursor will draw a line when you move it. You'll notice that, throughout the code, we use the `t.up()` and `t.down()` commands to pick the pen up when we want to move to a new location and put it down when we're ready to draw. We have some important steps to follow in this section, so let's get started:

1. Since the border of Colorado is a polygon, we must draw a line between the last point and the first point to close the polygon. We can also leave out the closing step and just add a duplicate point to the Colorado dataset. Once we've drawn the state, we'll use the `write()` method to label the polygon:

```python
t.up()
first_pixel = None
for point in state[POINTS]:
  pixel = convert(point)
  if not first_pixel:
    first_pixel = pixel
  t.goto(pixel)
  t.down()
t.goto(first_pixel)
t.up()
t.goto([0,0])
t.write(state[NAME], align="center", font=("Arial",16,"bold"))
```

2. If we were to run the code at this point, we would see a simplified map of the state of Colorado, as shown in the following screenshot:

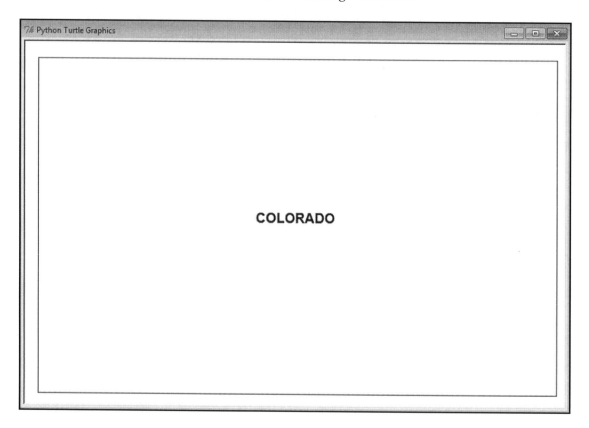

 If you do try to run the code, you'll need to temporarily add the following line at the end, or the Tkinter window will close as soon as it finishes drawing: `t.done()`.

3. Now, we'll render the cities as point locations and label them with their names and population. Since the cities are a group of features in a list, we'll loop through them to render them. Instead of drawing lines by moving the pen around, we'll use the turtle `dot()` method to plot a small circle at the pixel coordinate that's returned by our `SimpleGISconvert()` function. We'll then label the dot with the city's name and add the population. You'll notice that we must convert the population number into a string in order to use it in the turtle `write()` method. To do so, we will use Python's built-in `str()` function:

```
for city in cities:
    pixel = convert(city[POINTS])
    t.up()
    t.goto(pixel)
    # Place a point for the city
    t.dot(10)
    # Label the city
    t.write(city[NAME] + ", Pop.: " + str(city[POP]),
    align="left")
    t.up()
```

4. Now, we will perform one last operation to prove that we have created a real GIS. We will perform an attribute query on our data to determine which city has the largest population. Then, we'll perform a spatial query to see which city lies the furthest west. Finally, we'll print the answers to our questions on our thematic map page safely, out of the range of the map.

5. For our query engine, we'll use Python's built-in `min()` and `max()` functions. These functions take a list as an argument and return the minimum and maximum values of this list. These functions have a special feature called a key argument that allows you to sort complex objects. Since we are dealing with nested lists in our data model, we'll take advantage of the key argument in these functions. The key argument accepts a function that temporarily alters the list for evaluation before a final value is returned. In this case, we want to isolate the population values for comparison, and then the points. We could write a whole new function to return the specified value, but we can use Python's lambda keyword instead. The lambda keyword defines an anonymous function that is used inline. Other Python functions can be used inline, such as the string function, `str()`, but they are not anonymous. This temporary function will isolate our value of interest.

6. So, our first question is, *which city has the largest population?*

```
biggest_city = max(cities, key=lambda city:city[POP])
t.goto(0,-200)
t.write("The biggest city is: " + biggest_city[NAME])
```

7. The next question is, *which city lies the furthest west?*

```
western_city = min(cities, key=lambda city:city[POINTS])
t.goto(0,-220)
t.write("The western-most city is: " + western_city[NAME])
```

8. In the preceding query, we use Python's built-in `min()` function to select the smallest longitude value. This works because we represented our city locations as longitude and latitude pairs. It is possible to use different representations for points, including possible representations where this code would need modification to work correctly. However, for our `SimpleGIS`, we are using a common point representation to make it as intuitive as possible.

9. These last two commands are just for cleanup purposes. First, we hide the cursor. Then, we call the turtle `done()` method, which will keep the turtle graphics window with our map on it open until we choose to close it using the close handle at the top of the window:

```
t.pen(shown=False)
t.done()
```

10. Whether you followed along using the Python interpreter or you ran the complete program as a script, you should see the following map being rendered in real time:

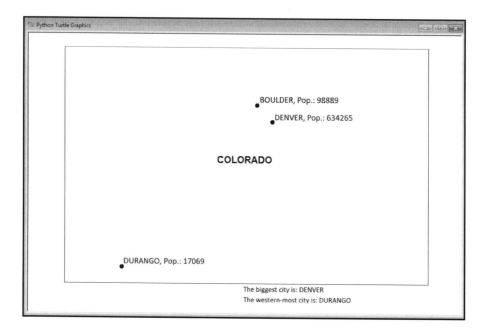

Congratulations! You have followed in the footsteps of Paleolithic hunters, the father of GIS Dr. Roger Tomlinson, geospatial pioneer Howard Fisher, and game-changing humanitarian programmers to create a functional, extensible, and technically complete geographic information system.

It took less than 60 lines of pure Python code! You will be hard-pressed to find a programming language that can create a complete GIS using only its core libraries in such a finite amount of readable code as Python. Even if you did, it is highly unlikely that the language would survive the geospatial Python journey that you'll take through the rest of this book.

As you can see, there is lots of room for expansion when it comes to SimpleGIS. Here are some other ways that you might expand this simple tool using the reference material for Tkinter and Python that were linked at the beginning of this section:

- Create an overview map in the top-right corner with a US border outline and Colorado's location in the US
- Add color for visual appeal and further clarity
- Create a map key for different features
- Make a list of states and cities and add more states and cities
- Add a title to the map
- Create a bar chart to compare population numbers visually

The possibilities are endless. SimpleGIS can also be used as a way to quickly test and visualize geospatial algorithms that you come across. If you want to add more data layers, you can create more lists, but these lists will become difficult to manage. In this case, you can use another Python module that's included in the standard distribution. The SQLite module provides a SQL-like database in Python that can be saved to disk or run in memory.

Summary

Well done! You are now a geospatial analyst. In this chapter, you learned about the history of geospatial analysis and the technologies that support it. You saw how Dr. Sarah Parcak's research made a big difference to history. You also became familiar with foundational GIS and remote sensing concepts that will serve you through the rest of this book. Finally, you took all of this knowledge and built a working GIS that can be expanded to do whatever you can imagine!

In the next chapter, we'll tackle the data formats that you'll encounter as a geospatial analyst. Geospatial analysts spend far more time dealing with data than actually performing analysis. Understanding the data that you're working with is essential to working efficiently and having fun.

Further reading

Here is a list of references you may refer to:

- If your Python distribution does not have Tkinter, you can find information on installing it from the following page: `https://tkdocs.com/tutorial/install.html`
- The official Python wiki page for Tkinter can be found here: `https://wiki.python.org/moin/TkInter`.
- The documentation for Tkinter is in the Python standard library documentation, which can be found at `https://docs.python.org/2/library/tkinter.html`.
- If you are new to Python, *Diveintopython* by Mark Pilgrim, published by Apress, is a free online book that covers all the basics of Python and will bring you up to speed. For more information, refer to `http://www.diveintopython.net`.

Learning Geospatial Data 2

One of the most challenging aspects of geospatial analysis is the data. Geospatial data already includes dozens of file formats and database structures and continues to evolve and grow to include new types of data and standards. Additionally, almost any file format can technically contain geospatial information and is done by simply adding a location.

In this chapter, we will look at the following topics:

- Getting an overview of common data formats
- Examining some common traits of geospatial data
- Understanding spatial indexing
- Knowing the most widely used vector data types
- Understanding raster data types

We'll also gain some insight into newer, more complex types, including point cloud data, web services, and geospatial databases.

Getting an overview of common data formats

As a geospatial analyst, you may frequently encounter the following general data types:

- Spreadsheets and **comma-separated values (CSV files)** or **tab-separated values (TSV files)**
- Geotagged photos
- Lightweight binary points, lines, and polygons
- Multi-gigabyte satellite or aerial images
- Elevation data such as grids, point clouds, or integer-based images
- XML files

- JSON files
- Databases (both servers and file databases)
- Web services
- Geodatabases

Each format contains its own challenges for access and processing. When you perform analysis on data, you usually have to do some form of preprocessing first. You might clip or subset a satellite image of a large area down to just your area of interest, or you might reduce the number of points in a collection to just the ones meeting certain criteria in your data model. A good example of this type of preprocessing is the `SimpleGIS` example that we looked at the end of `Chapter 1`, *Learning about Geospatial Analysis with Python*. The state dataset included just the state of Colorado rather than all 50 states. The city dataset included only three sample cities demonstrating three levels of population, along with different relative locations.

The common geospatial operations in `Chapter 1`, *Learning about Geospatial Analysis with Python*, are the building blocks for this type of preprocessing. However, it is important to note that there has been a gradual shift in the field of geospatial analysis toward readily available basemaps. Until around 2004, geospatial data was difficult to acquire and desktop computing power was much less than it is today. Preprocessing data was an absolute first step for any geospatial project. However, in 2004, Google released Google Maps, which wasn't long after Google Earth. Microsoft had also been developing a technology acquisition called **TerraServer**, which they relaunched around this time. In 2004, the **Open Geospatial Consortium** (**OGC**) updated the version of its **Web Map Service** (**WMS**), which was growing in use and popularity. This same year, Esri also released version 9 of its ArcGIS server system. These innovations were driven by Google's web map tiling model, which allowed for smooth, global, scrolling maps at many different resolutions, and were often called **slippy maps**.

People used map servers on the internet before Google Maps, most famously with the MapQuest driving directions website. However, these map servers offered only small amounts of data at a time and usually over limited areas. The Google tiling system converted global maps into tiered image tiles for both images and mapping data. These were served dynamically using JavaScript and the browser-based `XMLHttpRequest` API, more commonly known as **Asynchronous JavaScript and XML** (**AJAX**). Google's system scaled to millions of users using ordinary web browsers. More importantly, it allowed programmers to leverage JavaScript programming to create mashups so that they could use the Google Maps JavaScript API to add additional data to the maps. The mashup concept is actually a shared geospatial layers system. Users can combine and recombine data from different web services into a single map, as long as the data is web-accessible. Other commercial and open source systems quickly mimicked this concept.

Notable examples of distributed geospatial layers are **OpenLayers**, which provide an open source, Google-like API that has now gone beyond Google's API, offering additional features. Complimentary to OpenLayers is **OpenStreetMap**, which is the open source answer to the tiled map services consumed by systems such as OpenLayers. OpenStreetMap has global, street-level vector data and other spatial features that have been collected from available government data sources and the contributions of thousands of editors worldwide. OpenStreetMap's data maintenance model is similar to the way Wikipedia, the online encyclopedia, crowd sources information creation and updates for articles. Recently, even more mapping APIs have appeared, including Leaflet and Mapbox, which continue to increase in flexibility, simplicity, and capability.

The mashup revolution had interesting and beneficial side effects on data. Geospatial data is traditionally difficult to obtain. The cost of collecting, processing, and distributing data kept geospatial analysis constrained to those who could afford this steep overhead cost by producing data or purchasing it. For decades, geospatial analysis was the tool of governments, very large organizations, and universities. Once the web mapping trend shifted to large-scale, globally tiled maps, organizations began essentially providing basemap layers for free in order to draw developers to their platform. The massively scalable global map system required massively scalable, high-resolution data to be useful. Geospatial software producers and data providers wanted to maintain their market share and kept up with the technology trend.

Geospatial analysts benefited greatly from this market shift in several ways. First of all, data providers began distributing data in a common projection called **Mercator**. The Mercator projection is a nautical navigation projection that was introduced over 400 years ago. As we mentioned in `Chapter 1`, *Learning about Geospatial Analysis with Python*, all projections have practical benefits, as well as distortions. The distortion in the Mercator projection is its size. In a global view, Greenland appears bigger than the continent of South America. However, like every projection, it also has some benefits. Mercator preserves angles. Predictable angles allowed medieval navigators to draw straight bearing lines when plotting a course across oceans. Google Maps didn't launch with Mercator. However, it quickly became clear that roads in high and low latitudes met at odd angles on the map instead of the 90 degrees in reality.

Since the primary purpose of Google Maps was street-level driving directions, Google sacrificed the global view accuracy for far better relative accuracy among streets when viewing a single city. Competing mapping systems followed suit. Google also standardized on the WGS 84 datum. This datum defines a specific spherical model of the Earth, called a **geoid**. This model defines the normal sea level. What's significant about this choice by Google is that the **Global Positioning System** (**GPS**) also uses this datum. Therefore, most GPS units default to this datum as well, making Google Maps easily compatible with raw GIS data.

The Google variation of the Mercator projection is often called **Google Mercator**. The **European Petroleum Survey Group** (**EPSG**) assigns short numeric codes to projections as an easy way to reference them. Rather than waiting for the EPSG to approve or assign a code that was first only relevant to Google, they began calling the projection EPSG:900913, which is *Google* spelled with numbers. Later, EPSG assigned the code EPSG:3857, deprecating the older code. Most GIS systems recognize the two codes as synonymous. It should be noted that Google tweaked the standard Mercator projection slightly for its use; however, this variation is almost imperceptible. Google uses spherical formulas at all map scales, while the standard Mercator assumes an ellipsoidal form at large scales.

The following image of the Mercator projection (`https://en.wikipedia.org/wiki/File:Tissot_mercator.png`) was taken from Wikipedia:

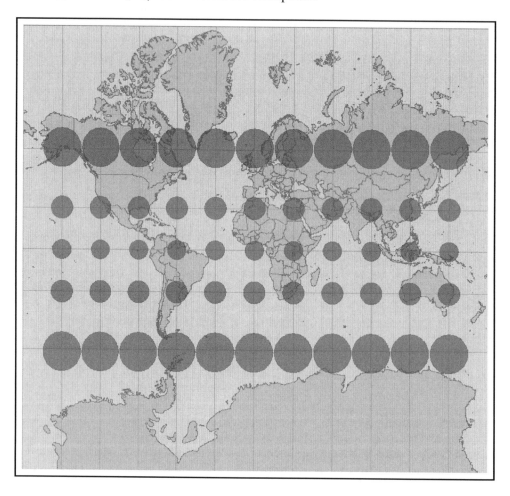

It shows the distortion caused by the Mercator projection using Tissot's indicatrix, which projects small ellipses of equal size on a map. The distortion of the ellipse clearly shows how the projection affects the size and distance: web mapping services have reduced the chore of hunting for data and much of the preprocessing for analysts to create basemaps. However, to create anything of value, you must understand geospatial data and how to work with it. This chapter provides an overview of the common data types and issues that you will encounter in geospatial analysis.

Throughout this chapter, two terms will be commonly used:

- **Vector data:** Vector data includes any format that minimally represents geolocation data using points, lines, or polygons.
- **Raster data:** Raster data includes any format that stores data in a grid of rows and columns. Raster data includes all image formats.

These are the two primary categories under which most geospatial datasets can be grouped.

If you want to see a projection that shows the relative size of continents more accurately, refer to the Goode homolosine projection: `https://en.wikipedia.org/wiki/Goode_homolosine_projection`.

Understanding data structures

Despite having dozens of formats, geospatial data has some common traits. Understanding these traits can help you approach and understand unfamiliar data formats by identifying the ingredients common to nearly all spatial data. The structure of a given data format is usually driven by its intended use.

Some data is optimized for efficient storage or compression, some is optimized for efficient access, some is designed to be lightweight and readable (web formats), while other data formats seek to contain as many different data types as possible.

Interestingly, some of the most popular formats today are also some of the simplest and even lack features found in more capable and sophisticated formats. Ease of use is extremely important to geospatial analysts because so much time is spent integrating data into geographic information systems, as well as exchanging data among analysts. Simple data formats facilitate these activities the best.

Common traits

Geospatial analysis is an approach in which you apply information processing techniques to data with a geographic context. This definition contains the most important elements of geospatial data:

- **Geolocation data**: Geolocation information can be as simple as a single point on the Earth referencing where a photo was taken. It can also be as complex as a satellite camera engineering model and orbital mechanics information being used to reconstruct the exact conditions and location under which the satellite captured the image.

- **Subject information**: Subject information can also cover a wide range of possibilities. Sometimes, the pixels in an image are the data in terms of a visual representation of the ground. Other times, an image may be processed using multispectral bands, such as infrared light, to provide information that's not visible in the image. Processed images are often classified using a structured color palette that is linked to a key, describing the information each color represents. Other possibilities include some form of database with rows and columns of information for each geolocated feature, such as the population associated with each city in our `SimpleGIS` from `Chapter 1`, *Learning about Geospatial Analysis with Python*.

These two factors are present in every format that can be considered geospatial data. Another common feature of geospatial data is spatial indexing. Overview datasets are also related to indexing.

Understanding spatial indexing

Geospatial datasets are often very large files, easily reaching hundreds of megabytes or even several gigabytes in size. Geospatial software can be quite slow in trying to repeatedly access large files when performing analysis.

As discussed briefly in `Chapter 1`, *Learning about Geospatial Analysis with Python*, spatial indexing creates a guide, which allows the software to quickly locate query results without examining every single feature in the dataset. Spatial indexes allow the software to eliminate possibilities and perform more detailed searches or comparisons on a much smaller subset of the data.

Spatial indexing algorithms

Many spatial indexing algorithms are derivatives of well-established algorithms that have been used on non-spatial information for decades. The two most common spatial indexing algorithms are **Quadtree index** and **R-tree index**.

Quadtree index

The Quadtree algorithm actually represents a series of different algorithms based on a common theme. Each node in a Quadtree index contains four children. These child nodes are typically square or rectangular in shape. When a node contains a specified number of features and more features are added, the node splits.

The concept of dividing a space into nested squares speeds up spatial searches. The software must only handle five points at a time and use simple greater-than/less-than comparisons to check whether a point is inside a node. Quadtree indexes are most commonly found in file-based index formats.

The following diagram shows a point dataset sorted by a Quadtree algorithm. The black points are the actual dataset, while the boxes are the bounding boxes of the index. Note that none of the bounding boxes overlap. The diagram on the left shows the spatial representation of the index, while the diagram on the right shows the hierarchical relationship of a typical index, which is how spatial software sees the index and data.

This structure allows a spatial search algorithm to quickly eliminate possibilities when trying to locate one or more points in relation to some other set of features, as shown in the following diagram:

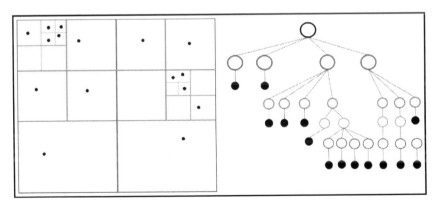

Now that we understand quadtree indexes, let's look at another common type of spatial indexes called R-trees.

R-tree index

R-tree indexes are more sophisticated than Quadtrees. R-trees are designed to handle 3D data and are optimized to store the index in a way that is compatible with the way databases use disk space and memory. Nearby objects are grouped together using an algorithm from a variety of spatial algorithms. All objects in a group are bounded by a minimum rectangle. These rectangles are aggregated into hierarchical nodes that are balanced at each level.

Unlike a Quadtree, the bounding boxes of an R-tree may overlap across nodes. Due to their relative complexity and database-oriented structure, R-trees are most commonly found in spatial databases, as opposed to file-based formats.

The following diagram from `https://en.wikipedia.org/wiki/File:R-tree.svg` shows a balanced R-tree for a 2D point dataset:

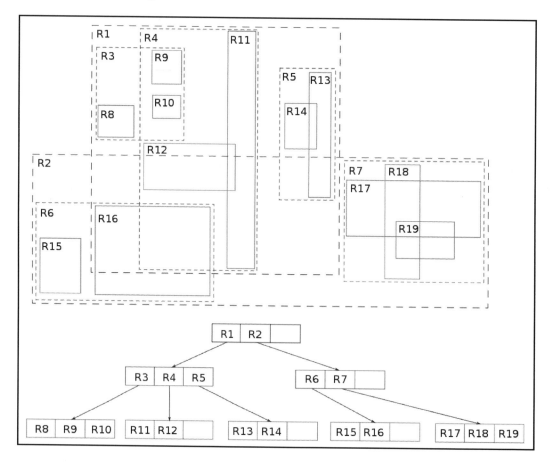

Indexes break up large datasets, but to speed up searching, they may employ a technique called **grids**. We'll look at that next.

Grids

Spatial indexes also often employ the concept of an integer grid. Geospatial coordinates are usually floating-point decimal numbers with anywhere from 2 to 16 decimal places. Performing comparisons on floating-point numbers is far more computationally expensive than working with integers. Indexed searching is about eliminating possibilities that don't require precision first.

Most spatial indexing algorithms, therefore, map floating-point coordinates to a fixed-sized integer grid. On searching for a particular feature, the software can use more efficient integer comparisons rather than working with floating-point numbers. Once the results are narrowed down, the software can access the full resolution data.

Grid sizes can be as small as 256 x 256 for simple file formats, or can be as large as 3 million x 3 million in large geospatial databases designed to incorporate every known coordinate system and possible resolution.

The integer mapping technique is very similar to the rendering technique that is used to plot data on a graphics canvas in mapping programs. The `SimpleGIS` script in `Chapter 1`, *Learning about Geospatial Analysis with Python*, also uses this technique to render points and polygons using the built-in Python turtle graphics engine.

What are overviews?

Overview data is most commonly found in raster formats. Overviews are resampled and lower-resolution versions of raster datasets that provide thumbnail views or simply faster-loading image views at different map scales. They are also known as **pyramids**, and the process of creating them is known as **pyramiding an image**. These overviews are usually preprocessed and stored with the full resolution data either embedded with the file or in a separate file.

The compromise of this convenience is that the additional images add to the overall file size of the dataset; however, they speed up image viewers. Vector data also has a concept of overviews, usually to give a dataset geographic context in an overview map. However, because vector data is scalable, reduced size overviews are usually created on the fly by software using a generalization operation, as mentioned in `Chapter 1`, *Learning about Geospatial Analysis with Python*.

Occasionally, vector data is rasterized by converting it into a thumbnail image, which is stored with, or embedded in, the image header. The following diagram demonstrates the concept of image overviews that shows visually why they are often called pyramids:

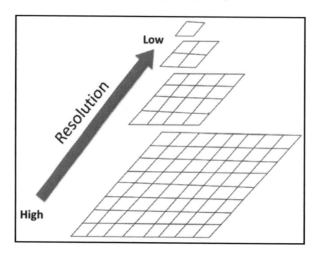

Spatial indexing and overviews help speed up access to data by software for analysts. Next, we'll look at metadata, which provides both a human-and machine-readable way to understand, search, and even catalog data.

What is metadata?

As discussed in Chapter 1, *Learning about Geospatial Analysis with Python*, metadata is any data that describes the associated dataset. Common examples of metadata include basic elements such as the footprint of the dataset on the Earth, as well as more detailed information such as spatial projection and information describing how the dataset was created.

Most data formats contain the footprint or bounding box of the data on the Earth. Detailed metadata is typically stored in a separate location in a standard format, such as the US **Federal Geographic Data Committee (FGDC)**, **Content Standard for Digital Geospatial Metadata (CSDGM)**, ISO, or the newer European Union initiative, which includes metadata requirements, and is called the **Infrastructure for Spatial Information in the European Community (INSPIRE)**.

Understanding the file structure

The preceding elements can be stored in a variety of ways in a single file, multiple files, or databases, depending on the format. Additionally, this geospatial information can be stored in a variety of formats, including embedded binary headers, XML, database tables, spreadsheets/CSV, separate text, or binary files.

Human-readable formats such as XML files, spreadsheets, and structured text files require only a text editor to be investigated. These files are also easily parsed and processed using Python's built-in modules, data types, and string manipulation functions. Binary-based formats are more complicated. Therefore, it is typically easier to use a third-party library to deal with binary formats.

However, you don't have to use a third-party library, especially if you just want to investigate the data at a high level. Python's built-in `struct` module has everything that you need. The `struct` module lets you read and write binary data as strings. When using the `struct` module, you need to be aware of the concept of byte order. Byte order refers to how the bytes of information that make up a file are stored in memory. This order is usually platform-specific, but in some rare cases, including shapefiles, the byte order is mixed into the file.

The Python `struct` module uses the greater than (>) and less than (<) symbols to specify byte order (big-endian and little-endian, respectively).

The following brief example demonstrates the usage of the Python `struct` module to parse the bounding box coordinates from an Esri shapefile vector dataset. You can download this shapefile as a zipped file from the following URL: `https://github.com/GeospatialPython/Learn/blob/master/hancock.zip?raw=true`.

When you unzip this, you will see three files. For this example, we'll be using `hancock.shp`. The Esri shapefile format has a fixed location and data type in the file header from byte 36 to byte 37 for the minimum x, minimum y, maximum x, and maximum y bounding box values. In this example, we will execute the following steps:

1. Import the `struct` module.
2. Open the `hancock.zip` shapefile in the binary read mode.
3. Navigate to byte 36.
4. Read each 8-byte double variables specified as d, and unpack it using the `struct` module in little-endian order, as designated by the < sign.

The best way to execute this script is in the interactive Python interpreter. We will then read the minimum longitude, minimum latitude, maximum longitude, and maximum latitude:

```
>>> import struct
>>> f = open("hancock.shp","rb")
>>> f.seek(36)
>>> struct.unpack("<d", f.read(8))
(-89.6904544701547,)
>>> struct.unpack("<d", f.read(8))
(30.173943486533133,)
>>> struct.unpack("<d", f.read(8))
(-89.32227546981174,)
>>> struct.unpack("<d", f.read(8))
(30.6483914869749,)
```

You'll notice that, when the struct module unpacks a value, it returns a Python tuple with one value. You can shorten the preceding unpacking code to one line by specifying all four doubles at once and increasing the byte length to 32 bytes, as shown in the following code:

```
>>> f.seek(36)
>>> struct.unpack("<dddd", f.read(32))
(-89.6904544701547, 30.173943486533133, -89.32227546981174,
    30.6483914869749)
```

Now that we understand how to describe data, let's learn about one of the most common types of geospatial data—vector data.

Knowing the most widely used vector data types

Vector data is, by far, the most common geospatial format because it is the most efficient way to store spatial information. In general, it requires fewer computer resources to store and process than raster data. The OGC has over 16 formats directly related to vector data. Vector data stores only geometric primitives, including points, lines, and polygons. However, only the points are stored for each type of shape. For example, in the case of a simple straight vector line shape, only the endpoints would be necessarily stored and defined as a line. Software displaying this data would read the shape type and then connect the endpoints with a line dynamically.

Geospatial vector data is similar to the concept of vector computer graphics, with some notable exceptions. Geospatial vector data contains positive and negative Earth-based coordinates, while vector graphics typically store computer screen coordinates. Geospatial vector data is also usually linked to other information about the object that's represented by the geometry. This information may be as simple as a timestamp in the case of GPS data or an entire database table for larger geographic information systems.

Vector graphics often store styling information describing colors, shadows, and other display-related instructions, while geospatial vector data typically does not. Another important difference is the shapes. Geospatial vectors typically only include very primitive geometries based on points, straight lines, and straight-line polygons, while many computer graphic vector formats have concepts of curves and circles. However, geospatial vectors can model these shapes using more points.

Other human-readable formats, such as CSV, simple text strings, GeoJSON, and XML-based formats, are technically vector data because they store geometry as opposed to rasters, which represent all the data within the bounding box of the dataset. Until the explosion of XML in the late 1990s, vector data formats were nearly all binary. XML provided a hybrid approach that was both computer and human-readable. The compromise is that text formats such as GeoJSON and XML data greatly increase the file size compared to binary formats. These formats will be discussed later in this section.

The number of vector formats to choose from is staggering. The open source vector library, OGR (`http://www.gdal.org/ogr_formats.html`), lists over 86 supported vector formats. Its commercial counterpart, Safe Software's **Feature Manipulation Engine** (**FME**), lists over 188 supported vector formats (`http://www.safe.com/fme/format-search/#filters%5B%5D=VECTOR`). These lists include a few vector graphics formats, as well as human-readable geospatial formats. There are still dozens of formats out there to at least be aware of, in case you come across them.

Now, let's look at a specific, and widely used type of vector data called shapefiles.

Shapefiles

The most ubiquitous geospatial format is the Esri shapefile. The geospatial software company known as Esri released the shapefile format specification as an open format in 1998 (`http://www.esri.com/library/whitepapers/pdfs/shapefile.pdf`). Esri developed it as a format for their ArcView software, designed as a lower-end GIS option to complement their high-end professional package, **ArcInfo**, formerly called **ARC/INFO**. However, the open specification, efficiency, and simplicity of the format turned it into an unofficial GIS standard that is still extremely popular over 15 years later.

Virtually, every piece of the software labeled as geospatial software supports shapefiles because the shapefile format is so common. For this reason, you can almost get by as an analyst by being intimately familiar with shapefiles and mostly ignoring other formats. You can convert almost any other format into a shapefile through the source format's native software or a third-party converter, such as the OGR library, for which there is a Python module. Other Python modules that handle shapefiles are Shapely and Fiona, which are based on OGR.

One of the most striking features of a shapefile is that the format consists of multiple files (from minimum to maximum, there can be 3-15 different files). The following table describes the file formats. The .shp, .shx, and .dbf files are required for a valid shapefile:

Shapefile supporting file extension	Supporting file purpose	Notes
.shp	This is the shapefile. It contains the geometry.	It is a required file. Some software that needs only geometry will accept the .shp files without the .shx or .dbf file.
.shx	This is the shape index file. It is a fixed-sized record index referencing geometry for faster access.	It is a required file. This file is meaningless without the .shp file.
.dbf	This is the database file. It contains the geometry attributes.	It is a required file. Some software will access this format without the .shp file present as the specification predates shapefiles. It's based on the very old FoxPro and dBase formats. An open specification exists for it called Xbase. The .dbf files can be opened by most types of spreadsheet software.
.sbn	This is the spatial bin file, that is, the shapefile spatial index.	It contains bounding boxes of features mapped to a 256 x 256 integer grid. It is very common for this file to accompany large shapefile datasets.
.sbx	A fixed-sized record index for the .sbn file.	A traditional ordered record index of a spatial index. Frequently seen.

Shapefile supporting file extension	Supporting file purpose	Notes
.prj	This contains map projection information that's stored in a well-known text format.	A very common and required file for on-the-fly projection by the GIS software. This same format can also accompany raster data.
.fbn	A spatial index of read-only features.	Very rarely seen.
.fbx	A fixed-sized record index of the .fbn spatial index.	Very rarely seen.
.ixs	A geocoding index.	Common in geocoding applications, including driving-direction type applications.
.mxs	Another type of geocoding index.	Less common than the .ixs format.
.ain	Attribute index.	Mostly legacy format, and rarely used in modern software.
.aih	Attribute index.	Accompanies the .ain files.
.qix	Quadtree index.	A spatial index format created by the open source community because the Esri .sbn and .sbx files were undocumented until recently.
.atx	Attribute index.	A more recent Esri software-specific attribute index to speed up attribute queries.
.shp.xml	Metadata.	A geospatial metadata .xml container. It can be any of the multiple XML standards, including FGDC and ISO.
.cpg	Code page file for .dbf.	It is used for the internationalization of .dbf files.

You will probably never encounter all of these formats at once. However, any shapefile that you use will have multiple files. You will commonly see .shp, .shx, .dbf, .prj, .sbn, .sbx, and occasionally, .shp.xml files. If you want to rename a shapefile, you must rename all of the associated files with the same name; however, in Esri software and other GIS packages, these datasets will appear as a single file.

Another important feature of shapefiles is that the records are not numbered. Records include the geometry, the .shx index record, and the .dbf record. These records are stored in a fixed order. When you examine shapefile records using the software, they appear to be numbered.

People are often confused when they delete a shapefile record, save the file, and reopen it; the number of the record that was deleted still appears. The reason for this is that the shapefile records are numbered dynamically on loading, but not saved. So, for example, if you delete record number 23 and save the shapefile, record number 24 will become 23 the next time you read the shapefile. Many people expect to open the shapefile and see the records jump from 22 to 24. The only way to track shapefile records in this way is to create a new attribute called ID or similar in the .dbf file and assign each record a permanent and unique identifier.

Just like renaming shapefiles, care must be taken while editing shapefiles. It's best to use software that treats the shapefiles as a single dataset. If you edit any of the files individually and add/delete a record without editing the accompanying files, the shapefile will be seen as corrupt by most geospatial software.

CAD files

CAD stands for **computer-aided design**. The primary formats for CAD data were created by Autodesk for their leading AutoCAD package. The two formats that are commonly seen are **Drawing Exchange Format (DXF)** and AutoCAD's native **Drawing (DWG)** format.

DWG was traditionally a closed format, but it has become more open.

The CAD software is used for everything that is engineering-related, from designing bicycles to cars, parks, and city sewer systems. As a geospatial analyst, you don't have to worry about mechanical engineering designs; however, civil engineering designs become quite an issue. Most engineering firms use geospatial analysis to a very limited degree, but store nearly all of their data in the CAD format. The DWG and DXF formats can represent objects using features not found in geospatial software or that are weakly supported by geospatial systems. Some examples of these features include the following:

- Curves
- Surfaces (for objects that are different from geospatial elevation surfaces)
- 3D solids
- Text (rendered as an object)
- Text styling
- Viewport configuration

These CAD and engineering-specific features make it difficult to cleanly convert CAD data into geospatial formats. If you encounter CAD data, the easiest option is to ask the data provider if they have shapefiles or some other geospatial-centric format.

Tag-based and markup-based formats

Tag-based markup formats are typically XML formats. They also include other structured text formats such as the **Well-Known Text (WKT)** format, which is used for projection information files as well as different types of data exchange.

XML formats include the **Keyhole Markup Language (KML)**, the **OpenStreetMap (OSM)** format, and the Garmin GPX format for GPS data, which has become a popular exchange format. The Open Geospatial Consortium's **Geographic Markup Language (GML)** standard is one of the oldest and most widely used XML-based geographic formats. It is also the basis for the OGC **Web Feature Service (WFS)** standard for web applications. However, GML has been largely superseded by KML and the GeoJSON format.

XML formats often contain more than just geometry. They also contain attributes and rendering instructions such as color, styling, and symbology. Google's KML format has become a fully supported OGC standard. The following is a sample of KML showing a simple placemark:

```
<?xml version="1.0" encoding="utf-8"?>
<kml xmlns="http://www.opengis.net/kml/2.2">
  <Placemark>
    <name>Mockingbird Cafe</name>
```

```
        <description>Coffee Shop</description>
        <Point>
          <coordinates>-89.329160,30.310964</coordinates>
        </Point>
      </Placemark>
    </kml>
```

The XML format is attractive to geospatial analysts for the following reasons:

- It is a human-readable format.
- It can be edited in a text editor.
- It is well supported by programming languages (especially Python).
- It is, by definition, easily extensible.

XML is not perfect, though. It is an inefficient storage mechanism for very large data formats and can quickly become cumbersome to edit. Errors in datasets are common, and most parsers do not handle errors robustly. Despite the downsides, XML is widely used in geospatial analysis.

Scalable Vector Graphics (**SVG**) is a widely supported XML format for computer graphics. It is supported well by browsers and is often used for geospatial rendering. However, SVG was not designed as a geographic format.

The WKT format is also an older OGC standard. The most common use for it is to define projection information usually stored in .prj projection files, along with a shapefile or raster. The WKT string for the WGS 84 coordinate system is as follows:

```
GEOGCS["WGS 84",
    DATUM["WGS_1984",
        SPHEROID["WGS 84",6378137,298.257223563,
            AUTHORITY["EPSG","7030"]],
        AUTHORITY["EPSG","6326"]],
    PRIMEM["Greenwich",0,
        AUTHORITY["EPSG","8901"]],
    UNIT["degree",0.01745329251994328,
        AUTHORITY["EPSG","9122"]],
    AUTHORITY["EPSG","4326"]]
```

The parameters that define a projection can be quite long. A standards committee, which was created by the EPSG, introduced a numerical coding system to reference projections. These codes, such as EPSG:4326, are used as shorthand for strings such as the preceding code. There are also short names for commonly used projections such as Mercator, which can be used in different software packages to reference a projection.

More information on these reference systems can be found on the spatial reference website at http://spatialreference.org/ref/.

GeoJSON

GeoJSON is a relatively new and brilliant text format based on the **JavaScript Object Notation (JSON)** format, which has been a commonly used data exchange format for years. Despite its short history, GeoJSON can be found embedded in all major geospatial software systems and most websites that distribute data. This is because JavaScript is the language of the dynamic web, and GeoJSON can be directly fed into JavaScript.

GeoJSON is a completely backward-compatible extension for the popular JSON format. The structure of JSON is very similar and, in some cases, identical to existing data structures of common programming languages. JSON is almost identical to Python's dictionary and list data types. Due to this similarity, parsing JSON in a script is simple to do from scratch, but there are many libraries to make it even easier. Python contains a built-in library aptly named json.

GeoJSON provides you with a standard way to define geometry, attributes, bounding boxes, and projection information. GeoJSON has all of the advantages of XML, including human-readable syntax, excellent software support, and wide use in the industry. It also surpasses XML.

GeoJSON is far more compact than XML, largely because it uses simple symbols to define objects rather than opening and closing text-laden tags. This compactness also helps with the readability and manageability of larger datasets. However, it is still inferior to binary formats from a data volume standpoint. The following is a sample of the GeoJSON syntax, defining a geometry collection with both a point and line:

```
{ "type": "GeometryCollection",
  "geometries": [
    { "type": "Point",
      "coordinates": [-89.33, 30.0]
    },
    { "type": "LineString",
      "coordinates": [ [-89.33, 30.30], [-89.36, 30.28] ]
    }
  {"type": "Polygon",
    "coordinates": [[
      [-104.05, 48.99],
      [-97.22,  48.98]
```

```
        }
    ]
}
```

The preceding code is a valid GeoJSON, but it is also a valid Python data structure. You can copy the preceding code sample directly into the Python interpreter as a variable definition and it will evaluate without error, as follows:

```
gc = { "type": "GeometryCollection",
    "geometries": [
        { "type": "Point",
          "coordinates": [-89.33, 30.0]
        },
        { "type": "LineString",
          "coordinates": [ [-89.33, 30.30], [-89.36, 30.28] ]
        }
    ]
}
gc
{'type': 'GeometryCollection', 'geometries': [{'type': 'Point',
    'coordinates': [
    -89.33, 30.0]}, {'type': 'LineString', 'coordinates': [[-89.33,
        30.3], [-89.36,30.28]]}]}
```

Due to its compact size, internet-friendly syntax by virtue of being written in JavaScript, and support from major programming languages, GeoJSON is a key component of leading REST geospatial web APIs, which will be covered later in this chapter. It currently offers the best compromise among the computer resource efficiency of binary formats, the human-readability of text formats, and programmatic utility.

GeoPackage

We'll briefly mention the GeoPackage format here as it's covered in Chapter 3, *The Geospatial Technology Landscape*, as well as because it's a type of geodatabase. The geopackage format is an OGC open standard on a SQLite file-based database container that is a platform, vendor, and software independent. It's an attempt to get away from all of the issues that are generated from either proprietary data formats or limited data formats.

Next, we'll look at the other major data type: raster data.

Understanding raster data types

Raster data consists of rows and columns of cells or pixels, with each cell representing a single value. The easiest way to think of raster data is as images, which is how they are typically represented by software. However, raster datasets are not necessarily stored as images. They can also be ASCII text files or **Binary Large Objects** (**BLOBs**) in databases.

Another difference between geospatial raster data and regular digital images is their resolution. Digital images express resolution as dots-per-inch if printed in full size. Resolution can also be expressed as the total number of pixels in the image, and are defined as megapixels. However, geospatial raster data uses the ground distance that each cell represents. For example, a raster dataset with a two-foot resolution means that a single cell represents two feet on the ground, which also means that only objects larger than two feet can be identified visually in the dataset.

Raster datasets may contain multiple bands, meaning that different wavelengths of light can be collected at the same time over the same area. Often, this range is from 3-7 bands, but can be several hundred in hyperspectral systems. These bands are viewed individually or swapped in and out as the RGB bands of an image. They can also be recombined into a derivative single-band image using mathematics and then recolored using a set number of classes representing values within the dataset.

Another common application of raster data is in the field of scientific computing, which shares many elements of geospatial remote sensing but adds some interesting twists. Scientific computing often uses complex raster formats, including **Network Common Data Form** (**NetCDF**), **GRIB**, and **HDF5**, which store entire data models. These formats are more like directories in a filesystem and can contain multiple datasets or multiple versions of the same dataset. Oceanography and meteorology are the most common applications of this kind of analysis. An example of a scientific computing dataset is the output of a weather model, where the cells of the raster dataset in different bands may represent a different variables' output from the model in a time series.

Like vector data, raster data can come in a variety of formats. The open source `raster` library, known as **Geospatial Data Abstraction Library** (**GDAL**), which actually includes the vector OGR library we mentioned earlier, lists over 130 supported raster formats (`http://www.gdal.org/formats_list.html`). The FME software package supports this many as well. However, just like shapefiles and CAD data, there are a few standout raster formats.

TIFF files

The **Tagged Image File Format (TIFF)** is the most common geospatial raster format. The TIFF format's flexible tagging system allows it to store any type of data whatsoever in a single file. TIFFs can contain overview images, multiple bands, integer elevation data, basic metadata, internal compression, and a variety of other data that's typically stored in additional supporting files by other formats. Anyone can extend the TIFF format unofficially by adding tagged data to the file structure. This extensibility has benefits and drawbacks. However, a TIFF file may work fine in one piece of software but fail when it's accessed in another because the two software packages implement the massive TIFF specification to different degrees. An old joke about TIFFs has a frustrating amount of truth to it: **TIFF** stands for **Thousands of Incompatible File Formats**. The GeoTIFF extension defines how geospatial data is stored. Geospatial rasters stored as TIFF files may have any of the following file extensions: `.tiff`, `.tif`, or `.gtif`.

JPEG, GIF, BMP, and PNG

JPEG, GIF, BMP, and PNG formats are common image formats in general, but can be used for basic geospatial data storage as well. Typically, these formats rely on accompanying the supporting text files for the georeferencing of the information in order to make them compatible with the GIS software, such as WKT, `.prj`, or world files.

The JPEG format is also fairly common for geospatial data. JPEGs have a built-in metadata tagging system, similar to TIFFs, called EXIF. JPEGs are commonly used for geotagged photographs in addition to raster GIS layers. **Bitmap (BMP)** images are used for desktop applications and document graphics. However, JPEG, GIF, and PNG are the formats that are used in web mapping applications, especially for pregenerated server map tiles for quick access via slippy maps.

Compressed formats

Since geospatial rasters tend to be very large, they are often stored using advanced compression techniques. The latest open standard is the JPEG 2000 format, which is an upgrade of the JPEG format and includes wavelet compression and a few other features, such as georeferencing data. The **Multi-resolution Seamless Image Database (MrSID)** (`.sid`) and **Enhanced Compression Wavelet (ECW)** (`.ecw`) are two proprietary wavelet compression formats often seen in geospatial contexts.

The TIFF format supports compression, including the **Lempel-Ziv-Welch** (**LZW**) algorithm. It must be noted that compressed data is suitable as part of a basemap, but should not be used for remote sensing processing. Compressed images are designed to look visually correct but often alter the original cell value. Lossless compression algorithms try to avoid degrading the source data, but it's generally considered a bad idea to attempt to perform spectral analysis on data that has been through compression. The JPEG format is designed to be a lossy format that sacrifices data for a smaller file size. It is also commonly encountered, so it is important to remember this fact to avoid invalid results.

ASCII Grids

Another means of storing raster data, often elevation data, is in ASCII Grid files. This file format was created by Esri, but has become an unofficial standard supported by most software packages. An ASCII Grid is a simple text file containing (x, y) values as rows and columns. The spatial information for the raster is contained in a simple header. The format of the file is as follows:

```
<NCOLS xxx>
<NROWS xxx>
<XLLCENTER xxx | XLLCORNER xxx>
<YLLCENTER xxx | YLLCORNER xxx>
<CELLSIZE xxx>
{NODATA_VALUE xxx}
row 1
row 2
  .
  .
  .
row n
```

While not the most efficient way to store data, ASCII Grid files are very popular because they don't require any special data libraries to create or access geospatial raster data. These files are often distributed as `.zip` files. The header values in the preceding format contain the following information:

- The number of columns
- The number of rows
- The x-axis cell center coordinate | x-axis lower-left corner coordinate
- The y-axis cell center coordinate | y-axis lower-left corner coordinate
- The cell size in mapping units
- The no-data value (typically, 9,999)

World files

World files are simple text files that can provide geospatial referencing information to any image externally for file formats that typically have no native support for spatial information, including JPEG, GIF, PNG, and BMP. The world file is recognized by geospatial software due to its naming convention. The most common way to name a world file is by using the raster file name and then altering the extension to remove the middle letter and adding w at the end.

The following table shows some examples of raster images in different formats and the associated world file name based on the convention:

Raster file name	World file name
World.jpg	World.jgw
World.tif	World.tfw
World.bmp	World.bpw
World.png	World.pgw
World.gif	World.gfw

The structure of a world file is very simple. It is a six-line text file, as follows:

- Line 1: The cell size along the *x*-axis in ground units
- Line 2: The rotation on the *y*-axis
- Line 3: The rotation on the *x*-axis
- Line 4: The cell size along the *y*-axis in ground units
- Line 5: The center *x*-coordinate of the upper-left cell
- Line 6: The center *y*-coordinate of the upper-left cell

The following is an example of world file values:

```
15.0
0.0
0.0
-15.0
-89,38
45.0
```

The (x, y) coordinates and the (x, y) cell size contained in lines 1, 4, 5, and 6 allow you to calculate the coordinate of any cell or the distance across a set of cells. The rotation values are important for geospatial software because remotely sensed images are often rotated due to the data collection platform.

Rotating the images runs the risk of resampling the data and, therefore, data loss, so the rotation values allow the software to account for the distortion. The surrounding pixels outside the image are typically assigned a `no data` value and represented as the color black.

The following image, courtesy of the **U.S. Geological Survey (USGS)** from `https://viewer.nationalmap.gov/advanced-viewer/`, demonstrates image rotation, where the satellite collection path is oriented from southeast to northeast, but the underlying basemap is north:

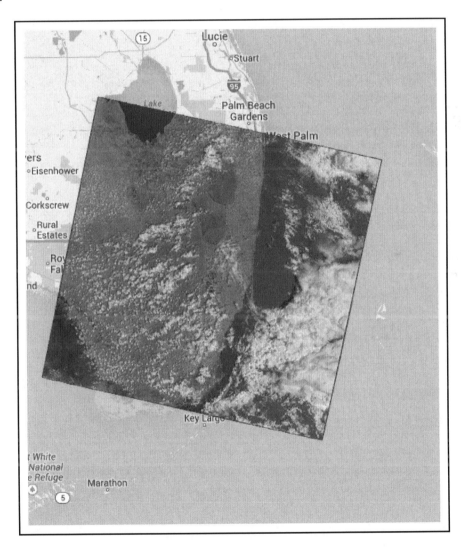

World files are a great tool when working with raster data in Python. Most geospatial software and data libraries support world files, so they are usually a good choice when it comes to georeferencing.

 You'll find that world files are very useful, but as you use them infrequently, you will forget what the unlabeled contents represent. A quick reference for world files is available at `https://kralidis.ca/gis/worldfile.htm`.

Vector data and raster data are the two most common data types. However, there is another type that is gaining popularity due to the cost of collecting it gradually becoming cheaper. That type is point cloud data, and we'll examine it next.

What is point cloud data?

Point cloud data is any data collected as the (x, y, z) location of a surface point based on some sort of focused energy return. This can be created using lasers, radar waves, acoustic soundings, or other waveform generation devices. The spacing between points is arbitrary and dependent on the type and position of the sensor collecting the data.

In this book, we will primarily be concerned with LIDAR data and radar data. Radar point cloud data is typically collected on space missions, while LIDAR is typically collected by terrestrial or airborne vehicles. Conceptually, both types of data are similar.

LIDAR

LIDAR uses powerful laser range-finding systems to model the world with very high precision. The term **LIDAR**, or LiDAR, is a combination of the words *light* and *radar*. Some people claim it also stands for **Light Detection and Ranging**. LIDAR sensors can be mounted on aerial platforms, including satellites, airplanes, or helicopters. They can also be mounted on vehicles for ground-based collection.

Due to the high-speed, continuous data collection provided by LIDAR, and a wide field of view – often 360 degrees of the sensor – LIDAR data doesn't typically have a rectangular footprint the way other forms of raster data do. LIDAR datasets are usually called point clouds because the data is a stream of (x,y,z) locations, where z is the distance from the laser to a detected object and the (x,y) values are the projected location of the object calculated from the location of the sensor.

The following image, courtesy of USGS, shows a point cloud LIDAR dataset in an urban area using a terrestrial sensor, as opposed to an aerial one. The colors are based on the strength of the laser's energy return, with red areas being closer to the LIDAR sensor and green areas farther away, which can give a precise height to within a few centimeters:

The most common data format for LIDAR data is the **LIDAR Exchange Format (LAS)**, which is a community standard. LIDAR data can be represented in many ways, including a simple text file with one (x, y, z) tuple per line. Sometimes, LIDAR data can be colorized using image pixel colors that have been collected at the same time. LIDAR data can also be used to create 2D elevation rasters.

This technique is the most common use for LIDAR in geospatial analysis. Any other use requires specialized software that allows the user to work in 3D. In that case, other geospatial data cannot be combined with the point cloud.

What are web services?

Geospatial web services allow users to perform data discovery, data visualization, and data access across the web. Web services are usually accessed by applications based on user input, such as zooming in to an online map or searching a data catalog. The most common protocols are the **Web Map Service (WMS)**, which returns a rendered map image, and **Web Feature Service (WFS)**, which typically returns GML, which was mentioned in this chapter's introduction.

Many WFS services can also return KML, JSON, zipped shapefiles, and other formats. These services are called through HTTP GET requests. The following URL is an example of a WMS GET request, which returns a map image of the world that is 640 pixels wide and 400 pixels tall and has an EPSG code of 900913: `http://ows.mundialis.de/services/` `service?SERVICE=wmsVERSION=1.1.1REQUEST=GetMapFORMAT=image/pngSTYLES=WIDTH=600` `HEIGHT=400LAYERS=TOPO-OSM-WMSSRS=EPSG:900913BBOX=-20037508,-` `20037508,20037508,20037508`.

Web services are rapidly evolving. The Open GIS Consortium is adding new standards for sensor networks and other geospatial contexts. **REpresentational State Transfer (REST)** services are also commonly used. REST services use simple URLs to make requesting data very easy to implement in nearly any programming language by tailoring URL parameters and their values accordingly. Nearly every programming language has robust HTTP client libraries that are capable of using REST services.

These REST services can return many types of data, including images, XML, or JSON. There is no overarching geospatial REST standard yet, but the OGC has been working on one for quite some time. Esri has created a working implementation that is currently widely used.

The following URL is an example of an Esri geospatial REST service that would return KML based on a weather radar image layer. You can add this URL to Google Earth as a network link, or you can download it as compressed KML (KMZ) in a browser to import it into another program: `https://idpgis.ncep.noaa.gov/arcgis/rest/services/NWS_Observatio` `ns/radar_base_reflectivity/MapServer/generateKml?docName=NWSRada` `r&layers=0&layerOptions=separateImage`.

You can find tutorials on the myriad of OGC services here: `http://cite.` `opengeospatial.org/pub/cite/files/edu/fundamental-concepts/text/` `basic.html`.

At the time of writing this book, the OGC is going through an API evolution that will significantly lower the barrier of using geospatial APIs through technologies such as REST, OpenAPI, JSON/HTML, and Swagger. You can track these trends through OGC's technology roadmap here: `https://github.com/opengeospatial/OGC-Technology-Trends`.

Now, we'll move from individual file formats to powerful geodatabases that can consolidate data through a single API.

Understanding geospatial databases

A geospatial database, or geodatabase, refers to an entire category of file formats, data schemas, and even software. In `Chapter 3`, *The Geospatial Technology Landscape*, we'll cover geodatabases as software packages, formally known as database management systems. But in this section, we'll describe their attributes as file formats. Geodatabases historically stored only vector data, though modern geodatabases are well-suited for raster data management as well.

Geodatabases can exhibit all of the common traits we noted previously. This information is stored in the database in what we call the database model. A very popular model is the traditional relational model, which uses tables of rows and columns. Each row and column combination is called a cell. Rows can be related to another table to link information using a designated column where each cell becomes a key referencing a cell in another table that then links the rows together.

The actual names of the columns and the relationships among data constitute the data definition. At a minimum, geodatabases associate a geometry description with attributes about the object the geometry represents. Single points are often represented by x and y columns. However, polygons and polylines have arbitrary numbers of points. This means that geodatabases often store geometry information as a **BLOB** using a format standard known as **Well-Known Binary**, or **WKB**.

The attribute information is usually defined as data types such as integers, floating-point decimal numbers, strings, or dates. The table may also include projection information for map display, as well as a spatial indexing column to speed up searching and geospatial comparisons. A geodatabase may also have another related table in order to link detailed metadata about the geospatial data.

Large geospatial raster datasets are rarely stored directly in the database. Typically, the raster data is stored on disk with a name, and a filesystem reference is stored in the database that points to the raster data. A geodatabase may also store a geometry column representing the ground footprint of the raster data, which can then be used as a proxy for geospatial operations.

Summary

You now have the background needed to work with common types of geospatial data. You also know about the common traits of geospatial datasets that will allow you to evaluate unfamiliar types of data and identify key elements that will drive you toward which tools to use when interacting with this data.

In the next chapter, we'll examine the modules and libraries that you can use to work with geospatial datasets. We will learn about the geospatial technology ecosystem, which consists of thousands of software libraries and packages. We will also understand the hierarchy of geospatial software and how it allows you to quickly comprehend and evaluate any geospatial tool.

Further reading

You can find tutorials on the myriad of OGC services here: `http://cite.opengeospatial.org/pub/cite/files/edu/fundamental-concepts/text/basic.html`.

3
The Geospatial Technology Landscape

The geospatial technology ecosystem consists of hundreds of software libraries and packages. This vast array of choices can be overwhelming for newcomers to geospatial analysis. The secret to learning geospatial analysis quickly is understanding the handful of libraries and packages that really matter. Most software, both commercial and open source, is derived from these critical packages. Understanding the ecosystem of geospatial software and how it's used allows you to quickly comprehend and evaluate any geospatial tool.

Geospatial libraries can be assigned to one or more of the following high-level core capabilities, which they implement to some degree. We will be learning about these capabilities in this chapter:

- Data access
- Computational geometry (including data reprojection)
- Image processing
- Visualization tools
- Metadata tools

In this chapter, we'll examine the packages that have had the largest impact on geospatial analysis, and also those that you are likely to frequently encounter. However, as with any filtering of information, you are encouraged to do your own research and draw your own conclusions.

The following websites offer more information on software that is not included in this chapter:

- Wikipedia list of GIS software: https://en.wikipedia.org/wiki/List_of_geographic_information_systems_software
- OSGeo project list and incubator projects: http://www.osgeo.org

The image processing software capability is for remote sensing. However, this category of software is very fragmented, containing dozens of software packages that are rarely integrated into derivative software. Most image processing software for remote sensing is based on the same data access libraries, with custom image processing algorithms implemented on top of them.

Take a look at the following examples of these types of software, which include both open source and commercial packages:

- **Open Source Software Image Map (OSSIM)**
- **Geographic Resources Analysis Support System (GRASS)**
- **Orfeo ToolBox (OTB)**
- **ERDAS IMAGINE**
- **ENVI**

Technical requirements

The following is a list of technical requirements for this chapter:

- Python 3.6 or higher
- RAM: Minimum 6 GB (Windows), 8 GB (macOS); recommended 8 GB
- Storage: Minimum 7200 RPM SATA with 20 GB of available space; recommended SSD with 40 GB of available space
- Processor: Minimum Intel Core i3 2.5 GHz; recommended Intel Core i5

Understanding data access

As described in `Chapter 2`, *Learning Geospatial Data*, geospatial datasets are typically large, complex, and varied. This challenge makes libraries that efficiently read, and in some cases write, this data essential to geospatial analysis. Without access to data, geospatial analysis cannot begin.

Furthermore, accuracy and precision are key factors in geospatial analysis. An image library that resamples data without permission, or a computational geometry library that rounds a coordinate by even a couple of decimal places, can adversely affect the quality of the analysis. Also, these libraries must manage memory efficiently. A complex geospatial process can last for hours, or even days.

If a data access library has a memory fault, it can delay an entire project or even an entire workflow, involving dozens of people who are dependent on the output of that analysis.

Data access libraries such as **Geospatial Data Abstraction Library (GDAL)** are mostly written in either C or C++ for speed and cross-platform compatibility. Speed is important due to the typically large size of geospatial datasets. However, you will also see many packages written in Java. When it's well written, pure Java can approach speeds that are acceptable for processing large vector or raster datasets, and that are usually acceptable for most applications.

The following concept map shows the major geospatial software libraries and packages and how they are related. The libraries in bold represent root libraries that are actively maintained, and are not significantly derived from any other libraries. These root libraries represent geospatial operations, which are rather difficult to implement, and the vast majority of people choose to use one of these libraries, rather than create a competing one. As you can see, a handful of libraries make up a disproportionate amount of geospatial analysis software. The following diagram is by no means exhaustive. In this book, we'll discuss only the most commonly used packages:

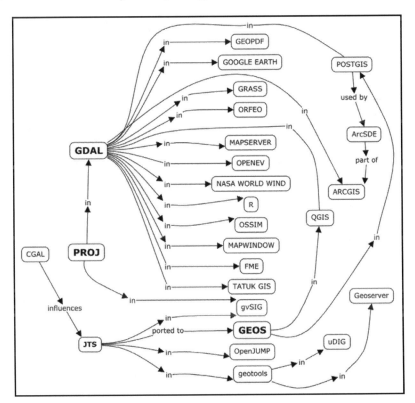

The **GDAL**, **GEOS** (short for **Geometry Engine - Open Source**), and **PROJ** libraries are the heart and soul of the geospatial analysis community on both the commercial and open source sides. It is important to note that these libraries are all written in C or C++. There is also significant work done in Java in the form of the **GeoTools** and **Java Topology Suite** (**JTS**) core libraries, which are used across a range of desktops, servers, and mobile software. Given that there are hundreds of geospatial packages available, with nearly all of them relying on these libraries to do anything meaningful, you'll beginning to get an idea of the complexity of geospatial data access and computational geometry. Compare this software domain to that of text editors, which return over 5,000 options when searched on the open source project site (`http://sourceforge.net/`).

Geospatial analysis is a truly worldwide community, with significant contributions to the field coming from every corner of the globe. But as you learn more about the heavy-hitting packages at the center of the software landscape, you'll see that these programs tend to come from Canada or are contributed to heavily by Canadian developers.

Credited as the birthplace of modern GIS, geospatial analysis is a matter of national pride. Also, the Canadian government and the public–private GeoConnections program have invested heavily in research and companies, both to fuel the industry for economic reasons, and out of necessity – to better manage the country's vast natural resources and the needs of its population.

GDAL

GDAL does the most heavy-lifting tasks in the geospatial industry. The GDAL website lists over 80 pieces of software using the library, and this list is by no means complete. Many of these packages are industry leading, open source, and commercial tools. This list doesn't include the hundreds of smaller projects and individual analysts who are using the library for geospatial analysis. GDAL also includes a set of command-line tools that can do a variety of operations without any programming.

A list of projects using GDAL can be found at the following URL: `http://trac.osgeo.org/gdal/wiki/SoftwareUsingGdal`.

GDAL and raster data

GDAL provides a single, abstract data model for the vast array of raster data types that are found in the geospatial industry. It consolidates unique data access libraries for different formats, and provides a common API for reading and writing data. Before the developer Frank Warmerdam created GDAL in the late 1990s, each data format required a separate data access library with a different API in order to read data, or in the worst-case scenario, developers often wrote custom data access routines.

The following diagram provides a visual description of how GDAL abstracts raster data:

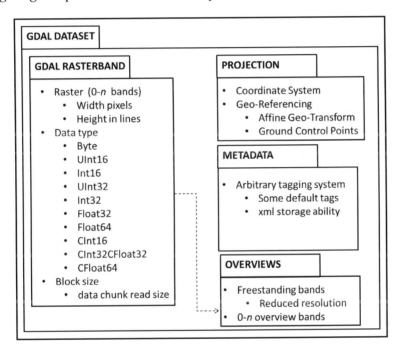

In the preceding software concept map, you can see that GDAL has had the greatest impact of any single piece of geospatial software. Combine GDAL with its sister library, OGR, for vector data, and the impact almost doubles. The PROJ library has also had a tremendous impact, but it is usually accessed via OGR or GDAL.

 The GDAL home page can be found at http://www.gdal.org/.

GDAL and vector data

In addition to raster data, GDAL lists at least partial support for over 70 vector data formats. Part of the success of the GDAL package is the X11/MIT open source license. This license is both commercial and open source-friendly. The GDAL library can be included in proprietary software without revealing the proprietary source code to users.

GDAL has the following vector capabilities:

- Uniform vector data and modeling abstraction
- Vector data reprojection
- Vector data format conversion
- Attribute data filtering
- Basic geometry filtering including clipping and point-in-polygon testing

GDAL has several command-line utility programs, which demonstrate its capability for vector data. This capability can also be accessed through its programming API. The following diagram outlines the GDAL vector architecture:

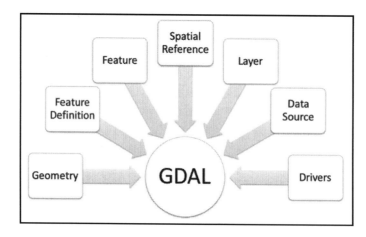

The GDAL vector architecture is fairly concise, considering this model is able to represent over 70 different data formats:

- **Geometry**: This object represents the **Open Geospatial Consortium (OGC)** Simple Features specification data model for points, linestrings, polygons, geometry collections, multipolygons, multipoints, and multilinestrings.
- **Feature Definition**: This object contains the attribute definitions of a group of related features.

- **Feature:** This object ties the **Geometry** and **Feature Definition** information together.
- **Spatial Reference**: This object contains an OGC spatial reference definition.
- **Layer**: This object represents features that are grouped as layers within a data source.
- **Data Source**: This object is the file or database object that is accessed by GDAL.
- **Drivers**: This object contains the translators for the 70-plus data formats that are available to GDAL.

This architecture works smoothly, with one minor quirk – the layer concept is used even for data formats that only contain a single layer. For example, shapefiles can only represent a single layer. But, when you access a shapefile using GDAL, you must still invoke a new layer object using the base name of the shapefile without a file extension. This design feature is only a minor inconvenience, heavily outweighed by the power that GDAL provides.

Now, let's go beyond accessing the data, to using it for analysis.

Understanding computational geometry

Computational geometry encompasses the algorithms that are needed to perform operations on vector data. The field is very old in computer science; however, most of the libraries used for geospatial operations are separate from computer graphics libraries because of geospatial coordinate systems. As described near the end of Chapter 1, *Learning about Geospatial Analysis with Python*, computer screen coordinates are almost always expressed in positive numbers, while geospatial coordinate systems often use negative numbers when they're moving west and south.

Several different geospatial libraries fit into this category, but they also serve a wide range of uses, from spatial selection to rendering. It should be noted that some features of GDAL that were described previously move it beyond the category of data access, and into the realm of computational geometry. But, it was included in the former category because that is its primary purpose.

Computational geometry is a fascinating subject. When writing a simple script to automate a geospatial operation, you inevitably need a spatial algorithm. The question then arises, *do you try to implement this algorithm yourself, or do you go through the overhead of using a third-party library?* The choice is always deceptive because some tasks are visually easy to understand and implement, some look complex but turn out to be easy, and some are trivial to comprehend but are extraordinarily difficult to implement. One such example is a geospatial buffer operation.

The concept is easy enough, but the algorithm turns out to be quite difficult. The following libraries in this section are the major packages that are used for computational geometry algorithms.

The PROJ projection library

U.S. Geological Survey (**USGS**) analyst, Jerry Evenden, created what is now known as the PROJ projection library in the mid-1990s while working at the USGS. Since then, it has become a project of the **Open Source Geospatial Foundation** (**OSGeo**), with contributions from many other developers. PROJ accomplishes the Herculean task of transforming data among thousands of coordinate systems. The math that is needed to convert points among so many coordinate systems is extremely complex. No other library comes close to the capability of PROJ. That fact and the routine that is needed by applications to convert datasets from different sources to a common projection make PROJ the undisputed leader in this area.

The following plot is an example of how specific the projections that are supported by PROJ can be. This plot from `https://calcofi.org` represents the line/station coordinate system of the **California Cooperative Oceanic Fisheries Investigations** (**CalCOFI**) program pseudo-projection, which is used only by **NOAA** (short for **National Oceanic and Atmospheric Administration**), the University of California Scripps Institution of Oceanography, and the California Department of Fish and Wildlife to collect oceanographic and fisheries data over the last 60 years along the California coastline:

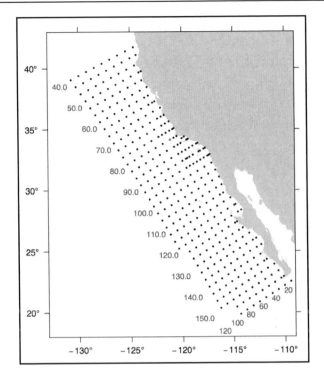

PROJ uses a simple syntax that is capable of describing any projection, including custom, localized ones, as shown in the previous plot. PROJ can be found in virtually every major GIS package, providing reprojection support, and it also has its own command-line tools.

It is available through GDAL for vector and raster data. However, it is often useful to access the library directly, because it gives you the ability to reproject individual points. Most of the libraries that incorporate PROJ only let you reproject entire datasets.

 For more information on PROJ, visit `https://proj4.org`.

CGAL

The **Computational Geometry Algorithms Library (CGAL)**, originally released in the late 1990s, is a robust and well-established open source computational geometry library. It was not specifically designed for geospatial analysis but is commonly used in the field.

CGAL is often referenced as a source for reliable geometry processing algorithms. The following diagram from the *CGAL User and Reference Manual* provides a visualization of one of the often-referenced algorithms from CGAL, called a polygon straight skeleton, which is needed to accurately grow or shrink a polygon:

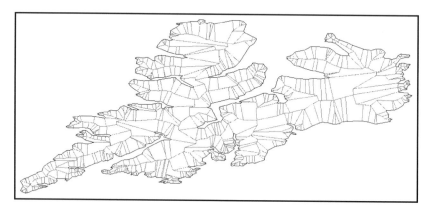

The straight skeleton algorithm is complex and important because shrinking or growing a polygon isn't just a matter of making it bigger or smaller. The polygon actually changes shape. As a polygon shrinks, non-adjacent edges collide and eliminate connecting edges. As a polygon grows, adjacent edges separate and new edges are formed to connect them. This process is key to the buffering of geospatial polygons. The following diagram, also from the *CGAL User and Reference Manual*, shows this effect using insets on the preceding polygon:

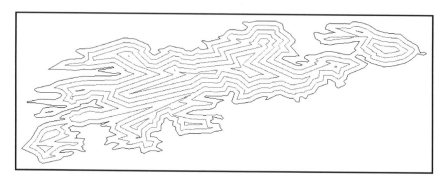

 CGAL can be found online at http://www.cgal.org/.

JTS

JTS is a geospatial computational geometry library that is written in 100% pure Java. JTS separates itself from other computational geometry libraries by implementing the OGC Simple Features specification for SQL. Interestingly, other developers have ported JTS to other languages, including C++, Microsoft .NET, and even JavaScript.

JTS includes a fantastic test program called the JTS TestBuilder, which provides a GUI to test functions without setting up an entire program. One of the most frustrating aspects of geospatial analysis concerns bizarre geometry shapes that break algorithms that work most of the time. Another common issue is unexpected results due to tiny errors in data such as polygons that intersect themselves in very small areas that are not easily visible. The JTS TestBuilder lets you interactively test JTS algorithms to verify data, or just to visually understand a process, as shown here:

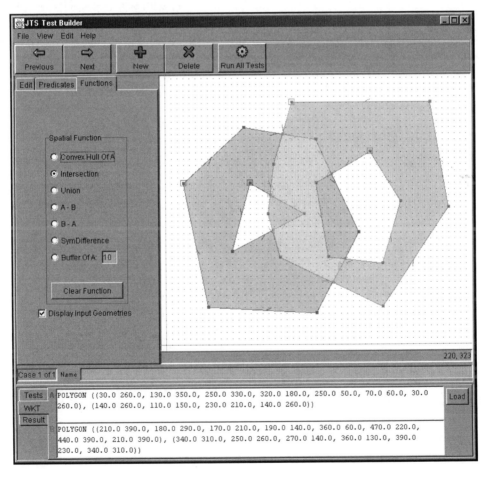

This tool is handy even if you aren't using JTS, but one of the several ports to another language. It should be noted that Vivid Solutions, the maintainer of JTS, hasn't released a new version since JTS Version 1.8 in December 2006. The package is quite stable and still in active use.

 The JTS home page is available at `https://locationtech.github.io/ jts`.

GEOS

GEOS is the C++ port of the JTS library that was explained previously. It is mentioned here because this port has had a much larger impact on geospatial analysis than the original JTS. The C++ version can be compiled on many platforms, as it avoids any platform-specific dependencies. Another factor responsible for the popularity of GEOS is that a fair amount of infrastructure exists to create automated or semi-automated bindings to various scripting languages, including Python. Another factor is that the majority of geospatial analysis software is written in C or C++. The most common use of GEOS is through other APIs that include GEOS as a library.

GEOS provides the following capabilities:

- OGC Simple Features
- Geospatial predicate functions
- Intersects
- Touches
- Disjoint
- Crosses
- Within
- Contains
- Overlaps
- Equals
- Covers
- Geospatial operations

- Union
- Distance
- Intersection
- Symmetric difference
- Convex hull
- Envelope
- Buffer
- Simplify
- Polygon assembly
- Polygon validation
- Area
- Length
- Spatial indexing
- OGC **Well-Known Text (WKT)** and **Well-Known Binary (WKB)** input/output
- C and C++ API
- Thread safety

GEOS can be compiled with GDAL to use all of its capabilities.

GEOS can be found online at `https://trac.osgeo.org/geos`.

PostGIS

As far as open source geospatial databases go, PostGIS is the most commonly used spatial database. PostGIS is essentially a module on top of the well-known PostgreSQL relational database. Much of the power of PostGIS comes from the previously mentioned GEOS library. Like JTS, it also implements the OGC Simple Features specification for SQL. This combination of computational geometry ability in a geospatial context sets PostGIS in a category of its own.

PostGIS allows you to execute both attribute and spatial queries against a dataset. Recall the point from Chapter 2, *Learning Geospatial Data*, that a typical spatial dataset is comprised of multiple data types including geometry, attributes (one or more columns of data in a row), and in most cases, indexing data. In PostGIS, you can query attribute data as you would do to any database table using SQL.

This capability is not surprising, as attribute data is stored in a traditional database structure. However, you can also query geometry using SQL syntax. Spatial operations are available through SQL functions, which you include as part of your queries. The following sample PostGIS SQL statement creates a 14.5 km buffer around the state of Florida:

```
SELECT ST_Buffer(the_geom, 14500)
FROM usa_states
WHERE state = 'Florida'
```

The FROM clause designates the usa_states layer as the location of the query. We filter that layer by isolating Florida in the WHERE clause. Florida is a value in the state column of the usa_states layer. The SELECT clause performs the actual spatial selection on the geometry of Florida that is normally contained in the the_geom column using the PostGIS ST_Buffer() function. The the_geom column is the geometry column for the PostGIS layer in this instance. The ST abbreviation in the function name stands for **spatial type**. The ST_Buffer() function accepts a column containing spatial geometries and a distance in the map units of the underlying layer.

The map units in the usa_states layer are expressed in meters, so 14.5 km would be 14,500 meters in the preceding example. Recall the point from Chapter 1, *Learning about Geospatial Analysis with Python,* that buffers like this query are used for proximity analysis. It just so happens that the Florida state water boundary expands 9 nautical miles, or approximately 14.5 km into the Gulf of Mexico from the state's western and northwestern coastlines.

The following map shows the official Florida state water boundary as a dotted line, which is labeled on the map:

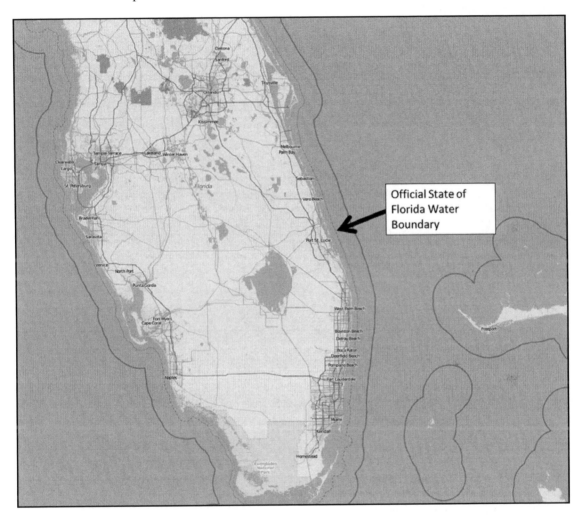

Official State of Florida Water Boundary

After applying the 9-nautical-mile buffer, you can see on the following map that the result, highlighted in orange, is quite close to the official legal boundary, which is based on detailed ground surveys:

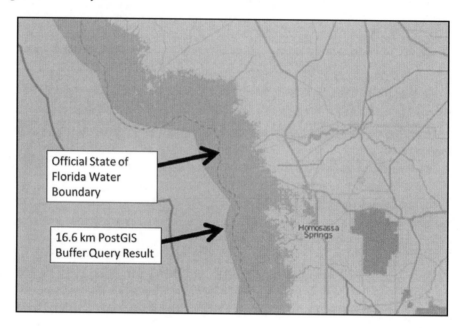

Currently, PostGIS maintains the following feature set:

- Geospatial geometry types, including points, linestrings, polygons, multipoints, multilinestrings, multipolygons, and geometry collections, which can store different types of geometries, including other collections of spatial functions for testing geometric relationships (for example, point-in-polygon or unions)
- Spatial functions for deriving new geometries (for example, buffers and intersects)
- Spatial measurements including perimeter, length, and area
- Spatial indexing using the R-tree algorithm
- A basic geospatial raster data type
- Topology data types
- U.S. geocoder based on **TIGER** (short for **Topologically Integrated Geographic Encoding and Referencing**) census data
- A new JSONB data type, which allows indexing and the querying of JSON and GeoJSON

The PostGIS feature set is competitive among all geodatabases, and is the most extensive among any open source or free geodatabases. The active momentum of the PostGIS development community is another reason for this system being best-of-breed. PostGIS is maintained at `http://postgis.net`.

Other spatially enabled databases

PostGIS is the gold standard among free and open source geospatial databases. However, there are several other systems that you should be aware of as a geospatial analyst. This list includes both commercial and open source systems, each with varying degrees of geospatial support.

Geodatabases have evolved in parallel with geospatial software, standards, and the web. The internet has driven the need for large, multiuser geospatial database servers that are able to serve large amounts of data. The following diagram, courtesy of `www.OSGeo.org`, shows how geospatial architectures have evolved, with a significant portion of this evolution happening at the database level:

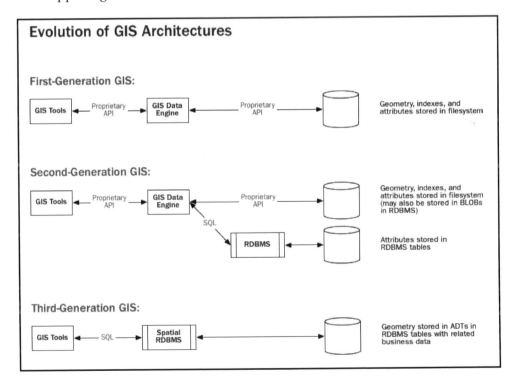

Oracle Spatial and Graph

The Oracle relational database is a widely used database system that is typically used by very large organizations because of its cost and large scalability. It is also extremely stable and fast. It runs some of the largest and most complicated databases in the world and is often found in hospitals, banks, and government agencies that manage millions of critical records.

Geospatial data capability first appeared in Oracle version 4 as a modification by the **Canadian Hydrographic Service (CHS)**. CHS also implemented Oracle's first spatial index, in the form of an unusual, but efficient, three-dimensional helical spiral. Oracle subsequently incorporated the modification and released the Oracle **Spatial Database Option (SDO)** in version 7 of the main database. The SDO system became Oracle Spatial in Oracle version 8. The database schema of Oracle Spatial still has the SDO prefix on some column and table names, similar to how PostGIS uses the OGC convention, ST, to separate spatial information from traditional relational database tables and functions at the schema level.

As of 2012, Oracle began calling the package Oracle Spatial and Graph, in order to emphasize the network data module. This module is used for analyzing networked datasets, such as transportation or utilities. However, the module can also be used against abstract networks, such as social networks. The analysis of social network data is a common target for big data analysis, which is now a growing trend. Big data social network analysis is likely the reason Oracle changed the name of the product.

Oracle Spatial has the following features:

- A geospatial data schema
- A spatial indexing system, which is now based on an R-tree index
- An SQL API for performing geometric operations
- A spatial data-tuning API to optimize a particular dataset
- A topology data model
- A network data model
- A GeoRaster data type to store, index, query, and retrieve raster data
- Three-dimensional data types, including **triangulated irregular networks (TINs)** and **LIDAR** (short for **Light Detection and Ranging**) point clouds
- A geocoder to search location names and return coordinates
- A routing engine for driving direction-type queries
- OGC compliance

Oracle Spatial and PostGIS are reasonably comparable and are both commonly used. You will see these two systems sooner or later as data sources while performing geospatial analysis.

Oracle Spatial and Graph is sold separately from Oracle itself. A little-known fact is that the SDO data type is native to the main Oracle database. If you have a simple application that just inputs points and retrieves them, you can use the main Oracle API to add, update, and retrieve SDOs without Oracle Spatial and Graph.

The U.S. **Bureau of Ocean Energy, Management, Regulation, and Enforcement (BOEMRE)** uses Oracle to manage environmental, business, and geospatial data for billions of dollars worth of oil, gas, and mineral rights, in one of the largest geospatial systems in the world. The following map is courtesy of BOEMRE:

Oracle Spatial and Graph can be found online at the following URL: `http://www.oracle.com/us/products/database/options/spatial/overview`.

ArcSDE

ArcSDE is Esri's **spatial data engine** (**SDE**). It is now rolled into Esri's ArcGIS Server product after over a decade of being a standalone product. What makes ArcSDE interesting is that the engine is mostly database-independent, supporting multiple database backends. ArcSDE supports IBM DB2, Informix, Microsoft SQL Server, Oracle, and PostgreSQL as data storage systems. While ArcSDE has the ability to create and manage a spatial schema from scratch on systems such as Microsoft SQL Server and Oracle, it uses native spatial engines if they are available. This arrangement is the case for IBM DB2, Oracle, and PostgreSQL. For Oracle, ArcSDE manages the table structure but can rely on the Oracle SDO data type for feature storage.

Like the previously mentioned geodatabases, ArcSDE also has a rich spatial selection API, and can handle raster data. However, ArcSDE does not have as rich a SQL spatial API as Oracle and PostGIS. Esri technically supports basic SQL functionality related to ArcSDE, but it encourages users and developers to use Esri software or programming APIs in order to manipulate data that is stored through ArcSDE, as it is designed to be a data source for Esri software.

Esri does provide software libraries to allow developers to build applications outside of Esri software using ArcSDE or Esri's file-based geodatabase, known as a personal geodatabase. But, these libraries are black boxes and the communication protocol that ArcSDE uses has never been reverse engineered. Typically, interaction happens between ArcSDE and third-party applications at the web services level using the ArcGIS Server API (which supports OGC services to some degree), and a fairly straightforward REST API service that returns GeoJSON.

The following screenshot is taken from the U.S. federal site, `http://catalog.data.gov`, a very large geospatial data catalog that is based on ArcSDE, which in turn networks U.S. federal data, including other ArcSDE installations from other federal agencies.

ArcSDE is integrated into ArcGIS Server; however, information on it can be found at `http://www.esri.com/software/arcgis/arcsde`.

Microsoft SQL Server

Microsoft added spatial data support to its flagship database product in Microsoft SQL Server 2008. It has gradually improved since that version but still is nowhere near as sophisticated as Oracle Spatial or PostGIS. Microsoft supports the same data types as PostGIS, but uses slightly different naming conventions, with the exception of rasters, which are not directly supported. It also supports output to WKT and WKB formats.

It offers some very basic support for spatial selection, but it is obviously not a priority for Microsoft at the moment. This limited support is likely because it is all that can be used for Microsoft software mapping components, and several third-party engines can provide spatial support on top of SQL Server.

Microsoft's support for spatial data in SQL Server is documented at the following
link: `http://msdn.microsoft.com/en-us/library/bb933790.aspx`.

MySQL

MySQL, another highly popular free database, provides nearly the same support as Microsoft SQL Server. The OGC geometry types are supported by basic spatial relationship functions. Through a series of buyouts, MySQL has become the property of Oracle.

While Oracle currently remains committed to MySQL as an open source database, this purchase has brought the ultimate future of the world's most popular open source database into question. But, as far as geospatial analysis is concerned, MySQL is barely a contender and is unlikely to be the first choice for any project.

For more information on MySQL spatial support, visit the following
link: `https://dev.mysql.com/doc/refman/8.0/en/spatial-types.html`

SpatiaLite

SpatiaLite is an extension of the open source SQLite database engine. SQLite uses a file database and is designed to be integrated into applications rather than into the typical client-server model, which is used by most relational database servers. SQLite already has spatial data types and spatial indexing, but SpatiaLite adds support for the OGC Simple Features specification, as well as map projections.

It should be noted that the extremely popular SQLite is not in the same category as Oracle, PostgreSQL, or MySQL, as it's a file-based database that is designed for single-user applications.

SpatiaLite can be found at `http://www.gaia-gis.it/gaia-sins/`.

GeoPackage

GeoPackage is a file-based geodatabase format. The official GeoPackage website, `http://geopackage.org`, describes it as:

> *"An open, standards-based, platform-independent, portable, self-describing, compact format for transferring geospatial information."*

It is also a direct answer to Esri's file geodatabase format, and also to the open geospatial community's designated *shapefile killer*, to replace the aging, partially closed shapefile format. Both formats are really file specifications, which rely on other software in order to read and write data.

GeoPackage is an OGC specification, which means its future is secure as an industry data format. It is also a *catch-all* format, which can handle vector data, raster data, attribute information, and extensions to address new requirements. And, like any good database, it handles multiple layers. You can store an entire GIS project in a single package, therefore making data management much simpler.

You can read more about Esri's file geodatabase format here: `http://desktop.arcgis.com/en/arcmap/10.3/manage-data/administer-file-gdbs/file-geodatabases.htm`.

Routing

Routing is a very niche area of computational geometry. It is also a very rich field of study that goes far beyond the familiar driving directions use case. The requirements for a routing algorithm are simply a networked dataset and impedance values that affect the speed of travel on that network. Typically, the dataset is vector-based, but raster data can also be used for certain applications.

The two major contenders in this area are Esri's Network Analyst, and the open source pgRouting engine for PostGIS. The most common routing problem is the most efficient way to visit a number of point locations. This problem is called the **traveling salesman problem (TSP)**. The TSP is one of the most intensely studied problems in computational geometry. It is often considered the benchmark for any routing algorithm.

More information on the TSP can be found at
`http://en.wikipedia.org/wiki/Travelling_salesman_problem`.

Esri Network Analyst and Spatial Analyst

Esri's entry into the routing arena, Network Analyst, is a truly generic routing engine that can tackle most routing applications regardless of context. Spatial Analyst is another Esri extension that is raster-focused, and it can perform *least-cost path* analysis on raster terrain data.

The ArcGIS Network Analyst product page is located on Esri's website
at: `http://www.esri.com/software/arcgis/extensions/networkanalyst`.

pgRouting

The **pgRouting** extension for PostGIS adds routing functionality to the geodatabase. It is oriented toward road networks, but it can be adapted to work with other types of networked data.

The following diagram shows a driving distance radius calculation output by pgRouting, which is displayed in QGIS. The points are color-coded from green to red, based on their proximity to the starting location. As shown in the following diagram, the points are nodes in the network dataset, courtesy of QGIS.org (`https://qgis.org/en/site/`), which in this case are roads:

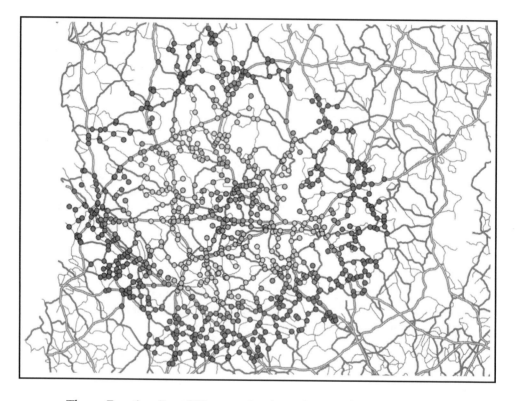

 The pgRouting PostGIS extension is maintained at
`http://pgrouting.org/`.

Next, we'll look at the tools that you need in order to visualize diagrams like the previous one.

Understanding desktop tools (including visualization)

Geospatial analysis requires the ability to visualize output. This fact makes tools that can visualize data absolutely critical to the field. There are two categories of geospatial visualization tools.

The first is geospatial viewers and the second is geospatial analysis software. The first category – geospatial viewers—allows you to access, query, and visualize data, but not to edit it in any way. The second category allows you to perform those tasks, and edit the data, too. The main advantage of viewers is that they are typically lightweight pieces of software that launch and load data quickly.

Geospatial analysis software requires far more resources to be able to edit complex geospatial data, so it loads more slowly and often renders data more slowly, in order to provide dynamic editing functionality.

Quantum GIS

Quantum GIS, more commonly known as **QGIS**, is a complete open source geographic information system. QGIS falls well within the geospatial analysis category in the two categories of visualization software. The development of the system began in 2002 and Version 1.0 was released in 2009.

It is the best showcase of most of the libraries that have been previously mentioned in this chapter. QGIS is written in C++, using the Qt library for the GUI. The GUI is well designed and easy to use. In fact, a geospatial analyst who has been trained on a proprietary package, such as Esri's ArcGIS or Manifold system, will be right at home using QGIS. The tools and menu system are logical and typical of a GIS system. The overall speed of QGIS is as good as, or better than, any other system that is available.

A nice feature of QGIS is that the underlying libraries and utility programs are just underneath the surface. Modules can be written by any third party in Python and added to the system. QGIS also has a robust online package management system to search for, install, and update these extensions. The Python integration includes a console that allows you to issue commands at the console and see the results in the GUI. QGIS isn't the only software to offer this capability.

Like most geospatial software packages, with Python integration, it installs a complete version of Python if you use the automated installer. There's no reason to worry if you already have Python installed. Having multiple versions of Python on a single machine is fairly common and well supported. Many people have multiple versions of Python on their computers for the purpose of testing software, or because it is such a common scripting environment for so many different software packages.

When the Python console is running in QGIS, the entire program API is available through an automatically loaded object called `qgis.utils.iface`. The following screenshot shows QGIS with the Python console running:

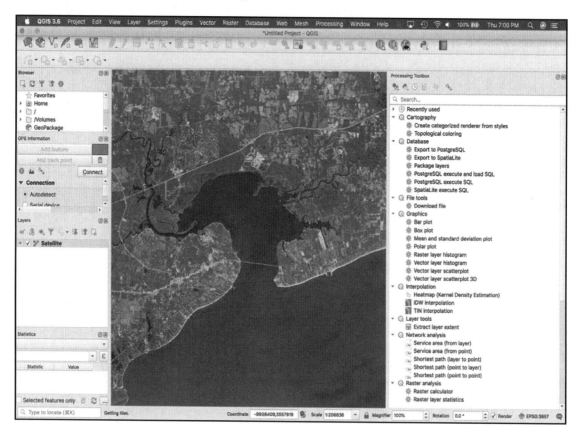

Because QIS is based on GDAL/OGR and GEOS, and it can use PostGIS, it supports all of the data sources that are offered by those packages. It also has nice raster processing features, too. QGIS works well for producing paper maps or entire map books using available extensions.

 QGIS is well documented on the QGIS website at the following link: `http://www.qgis.org/en/documentation.html`. You can also find numerous online and video tutorials by searching for QGIS or particular operation.

OpenEV

OpenEV is an open source geospatial viewer that was originally developed by Atlantis Scientific around 2002, which became Vexcel, before a buyout by Microsoft. Vexcel developed OpenEV as a freely downloadable satellite image viewer for the **Canadian Geospatial Data Infrastructure (CGDI)**. It is built using GDAL and Python and is partially maintained by GDAL creator, Frank Warmerdam.

OpenEV is one of the fastest raster viewers available. Despite being originally designed as a viewer, OpenEV offers all of the utility of GDAL and PROJ. While created as a raster tool, it can overlay vector data such as shapefiles, and even supports basic editing. Raster images can also be altered using the built-in raster calculator, and data formats can be converted, reprojected, and clipped.

The following screenshot shows a 25 MB, 16-bit integer GeoTIFF elevation file in an OpenEV viewer window:

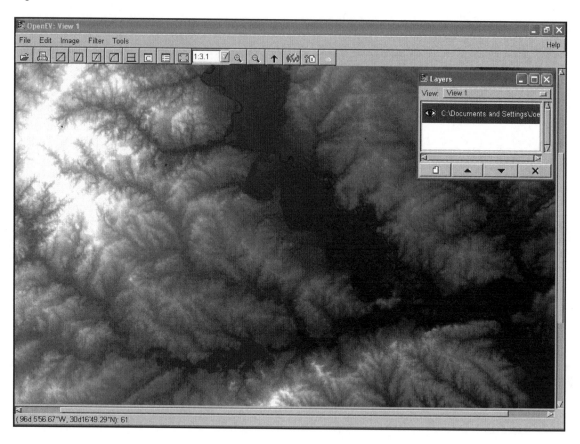

OpenEV is built largely in Python and offers a Python console with access to the full capability of the program. The OpenEV GUI isn't as sophisticated as other tools, such as QGIS. For example, you cannot drag and drop geospatial datasets into the viewer as you can do in QGIS. But, the raw speed of OpenEV makes it very attractive for simple raster viewing, or for basic processing and data conversion.

 The OpenEV home page is available at http://openev.sourceforge.net.

GRASS GIS

GRASS is one of the oldest continuously developed geospatial systems in existence. The U.S. Army Corps of Engineers began GRASS development in 1982. It was originally designed to run on Unix systems. In 1995, the army released the last patch, and the software was transferred to community development, where it has remained ever since.

Even though the user interface was redesigned, GRASS still feels somewhat esoteric to modern GIS users. However, because of its decades-old legacy and non-existent price tag, many geospatial workflows and highly specialized modules have been implemented in GRASS over the years, making it highly relevant to many organizations and individuals, especially in research communities. For these reasons, GRASS is still actively developed.

GRASS has also been integrated with QGIS, so the more modern and familiar QGIS GUI can be used to run GRASS functions. GRASS is also deeply integrated with Python and can be used as a library or as a command-line tool. The following screenshot shows some landform analysis in the native GRASS GUI, which was built using the `wxPython` library:

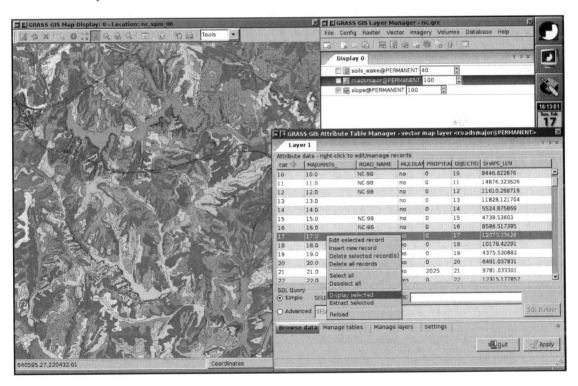

GRASS is housed online at `http://grass.osgeo.org/`.

gvSIG

Another Java-based desktop GIS is **gvSIG**. The gvSIG project began in 2004 as part of a larger project to migrate the IT systems of the Regional Ministry of Infrastructure and Transport of Valencia, Spain, to free software. The result was gvSIG, which has continued to mature. The feature set is mostly comparable to QGIS, with some unique capabilities as well.

The official gvSIG project has a very active fork called **gvSIG Community Edition (gvSIG CE)**. There is also a mobile version called gvSIG mobile. The gvSIG code base is open source.

The official home page for gvSIG is available at `http://www.gvsig.org/web/`.

OpenJUMP

OpenJUMP is another open source Java-based desktop GIS. **JUMP** stands for **Java Unified Mapping Platform**, and was originally created by Vivid Solutions for the Government of British Columbia. After Vivid Solutions delivered JUMP, development stopped. Vivid Solutions eventually released JUMP to the open source community, where it was renamed OpenJUMP.

OpenJUMP has the ability to read and write shapefiles and OGC **GML** (short for **Geography Markup Language**), and supports PostGIS databases. It can also display some image formats and data from OGC **WMS** (short for **Web Map Server**) and **WFS** (short for **Web Feature Service**) services. It has a plugin architecture, and it can also serve as a development platform for custom applications.

You can find out more about OpenJUMP on the official web page at
http://www.openjump.org/.

Google Earth

Google Earth is so ubiquitous that it hardly seems worth mentioning. The first release of EarthViewer 3D in 2001 (created by a company called Keyhole Inc.), and the EarthViewer 3D project were funded by the non-profit venture capital firm, In-Q-Tel, which in turn, is funded by the U.S. Central Intelligence agency. This spy agency lineage and the subsequent purchase of Keyhole by Google to create and distribute Google Earth brought global attention to the geospatial analysis field.

Since the first release of the software as Google Earth in 2005, Google has continually refined it. Some of the notable additions are the creation of Google Moon, Google Mars, Google Sky, and Google Oceans. These are virtual globe applications, which feature data from the Moon and Mars, with the exception of Google Oceans, which adds sea-floor elevation mapping, known as bathymetry, to Google Earth.

Google Earth introduced the idea of the spinning virtual globe concept for the exploration of geographic data. After centuries of looking at 2D maps, or low-resolution physical globes, flying around the Earth virtually and dropping onto a street corner anywhere in the world was mind-blowing – especially for geospatial analysts and other geography enthusiasts, as depicted in the following screenshot of Google Earth, overlooking the Central Business District in New Orleans, Louisiana:

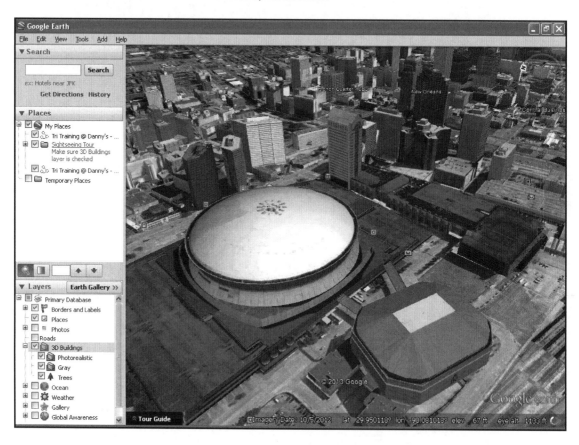

Just as Google had revolutionized web mapping with its tile-based *slippy* mapping approach, the virtual globe concept was a major boost to geospatial visualization.

After the initial excitement wore off, many geospatial analysts realized that Google Earth was a very animated and fun geographic exploration tool, but it really had a very limited utility for any kind of meaningful geospatial analysis. Google Earth falls squarely into the realm of geospatial viewer software.

The only data format it consumes is its native **keyhole markup language (KML)**, which is an all-in-one data and styling format, which is discussed in `Chapter 2`, *Learning Geospatial Data*. As this format is now an OGC standard, consuming only one data format immediately limits the utility of any tool. Any project involving Google Earth must first begin with complete data conversion and styling in KML, reminiscent of geospatial analysis from around 10-20 years ago. The tools that do support KML, including Google Maps, support a limited subset of KML.

Google Earth's native dataset has global coverage, but it is a mixture of datasets spanning several years and sources. Google has greatly improved the inline metadata in the tool, which identifies the source and approximate date of the current view. But, this method creates confusion among lay people. Many people believe that the data in Google Earth is updated far more frequently than it really is. The Google Street View system, showing street-level, 360-degree views of much of the world, has somewhat helped to correct this misperception.

People are able to easily identify images of familiar locations as being several years old. Another common misperception created by Google Earth is that the entire world has been mapped in detail, and therefore creating a base map for geospatial analysis should be trivial. As discussed in `Chapter 2`, *Learning Geospatial Data*, mapping an area of interest is far easier than it used to be a few years ago by using modern data and software, but it is still a complex and labor-intensive endeavor. This misperception is one of the first customer expectations a geospatial analyst must manage when starting a project.

Despite these misperceptions, the impact that Google has had on geospatial analysis is almost entirely positive. For decades, one of the most difficult challenges to growing the geospatial industry was convincing potential stakeholders that geospatial analysis is almost always the best approach when making decisions about people, resources, and the environment. This hurdle stands in sharp contrast to a car dealer. When a potential customer comes to a car lot, the salesman doesn't have to convince the buyer that they need a car, just about the type of car.

Geospatial analysts had to first educate project sponsors on the technology, and then convince them that the geospatial approach was the best way to address a challenge. Google has largely eliminated those steps for analysts.

 Google Earth can be found online at
`http://www.google.com/earth/index.html`.

NASA WorldWind

NASA WorldWind is an open source virtual globe and geospatial viewer, originally released by the U.S. **National Aeronautics and Space Administration (NASA)**, in 2004. It was originally based on Microsoft's .NET Framework, making it a Windows-centric application.

The following screenshot of NASA WorldWind looks similar to Google Earth:

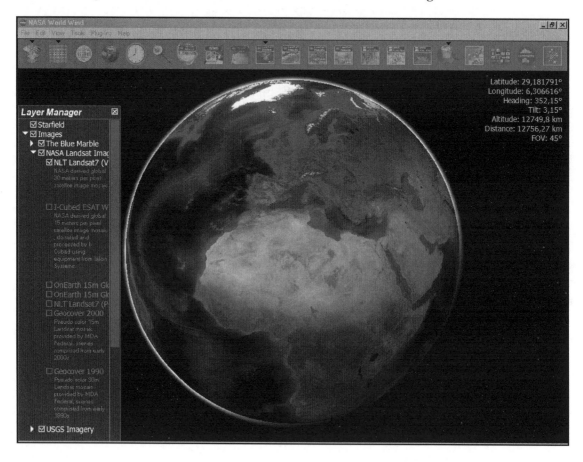

In 2007, a Java-based **software development kit (SDK)** was released, called **WorldWind Java**, which made WorldWind more cross-platform. The transition to Java also led to the creation of a browser plugin for WorldWind.

The WorldWind Java SDK is considered an SDK and not a desktop application like the .NET version. However, the demos included with the SDK provide a viewer without any additional development. While NASA WorldWind was originally inspired by Google Earth, its status as an open source project takes it in an entirely different direction.

Google Earth is a generalist tool that is bounded by the limits of the KML specification. NASA WorldWind is now a platform upon which anyone can develop without limits. As new types of data become available and computing resources grow, the potential of the virtual globe paradigm certainly holds more potential for geospatial visualization that has not yet been explored.

NASA WorldWind is available online at
`http://worldwind.arc.nasa.gov/java/`.

ArcGIS

Esri walks the line of being one of the greatest promoters of the geospatial analytical approach to understanding our world, and is a privately held, profit-making business, which must look out for its own interests to a certain degree. The ArcGIS software suite represents every type of geospatial visualization known, including vector, raster, globes, and 3D. It is also a market leader in many countries. As shown in the geospatial software map earlier in this chapter, Esri has increasingly incorporated open source software into its suite of tools, including GDAL for raster display, and Python as the scripting language for ArcGIS.

The following screenshot shows the core ArcGIS application, ArcMap, with marine tracking density data analysis. The interface shares a lot in common with QGIS, as shown in this screenshot courtesy of `https://marinecadastre.gov/`:

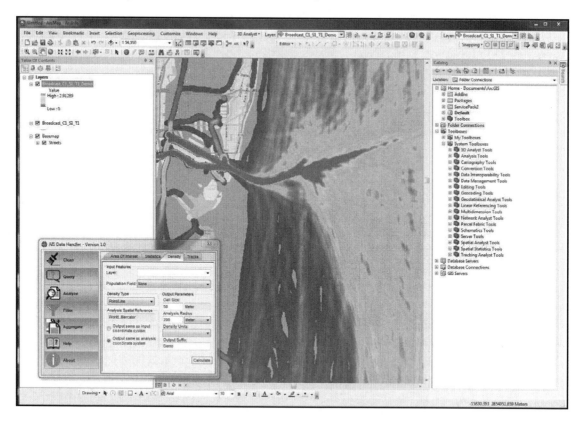

The ArcGIS product page is available online at `http://www.esri.com/software/arcgis`.

Now that we understand the tools for visualizing and analyzing data, let's look at how to manage data.

Understanding metadata management

The distribution of data on the internet has increased the importance of metadata. Data custodians are able to release a dataset to the entire world for download without any personal interaction. The metadata record of a geospatial dataset can follow this to help ensure that the integrity and accountability of this data are maintained.

Properly formatted metadata also allows for automated cataloging, search indexing, and the integration of datasets. Metadata has become so important that a common mantra within the geospatial community is *data without metadata isn't data*, meaning that a geospatial dataset cannot be fully utilized and understood without metadata.

The following section lists some of the common metadata tools that are available. The OGC standard for metadata management is the **Catalog Service for the Web (CSW)**, which creates a metadata-based catalog system and an API for distributing and discovering datasets.

Python's pycsw Library

pycsw is an OGC-compliant CSW for the publishing and discovery of geospatial metadata. It supports multiple APIs, including CSW 2/CSW 3, OpenSearch, OAI-PMH, and SRU. It is extremely lightweight and pure Python. For an excellent example of a CSW and client that was built using pycsw, see the **Pacific Islands Ocean Observing System (PacIOOS)** catalog, available at the following link: http://pacioos.org/search/. The pycsw library is also used in a larger package called **GeoNode**.

GeoNode

GeoNode is a Python-based geospatial content management system. It combines geospatial data creation, metadata, and visualization in a single server package. It also includes social features, such as comments and rating systems. It is open source and is available at `http://geonode.org/`. The following screenshot is from the GeoNode online demo:

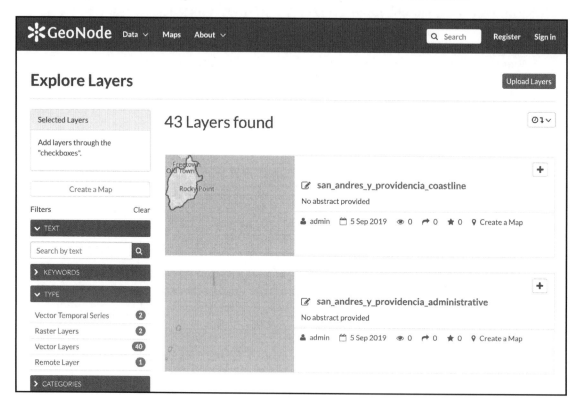

GeoNode and `pycsw` are the two main metadata tools for Python. Next, we'll look at some tools that are written in other languages.

GeoNetwork

GeoNetwork is an open source, Java-based catalog server used to manage geospatial data. It includes a metadata editor and search engine, as well as an interactive web map viewer. The system is designed to globally connect spatial data infrastructures. It can publish metadata through the web using the metadata editing tools. It can also publish spatial data through the embedded GeoServer map server. It has user and group security permissions, and web and desktop configuration utilities.

GeoNetwork can also be configured in order to harvest metadata from other catalogs at scheduled intervals. The following screenshot is of the United Nations Food and Agriculture Organization's implementation of GeoNetwork:

You can find out more about GeoNetwork at
`http://geonetwork-opensource.org/`.

Summary

In this chapter, you learned about the hierarchy of geospatial analysis software. You also learned a framework for approaching the hundreds of existing geospatial software packages and libraries, by categorizing them into one or more major functions, including data access, computational geometry, raster processing, visualization, and metadata management.

We also examined commonly used foundation libraries, including GDAL, OGR, PROJ, and GEOS, which are found again and again in geospatial software. You can approach any new piece of geospatial software, trace it back to these core libraries, and then ask, *what is the value added?* to gain a better understanding of the package. If the software isn't using one of these libraries, you need to ask, *why are these developers going against the grain?* in order to understand what that system brings to the table.

Python was only mentioned a few times in this chapter so as to avoid any distraction in understanding the geospatial software landscape. But, as we will see, Python is interwoven into every single piece of software in this chapter and is a fully capable geospatial tool in its own right. It is no coincidence that Python is the official scripting language of ArcGIS, QGIS, GRASS, and many other packages. It is also not by chance that GDAL, OGR, PROJ, CGAL, JTS, GEOS, and PostGIS all have Python bindings.

And as for the packages not mentioned here, they are all within Python's grasp through the Jython Java distribution, the IronPython .NET distribution, Python's various database and web APIs, and the built-in `ctypes` module. As a geospatial analyst, if there's one technology you can't afford to pass up, it's Python.

In the next chapter, we'll see how Python comes into the picture in the geospatial industry. We'll also learn about the GIS scripting language, the mashup glue language, and the full-blown programming language.

Further reading

Here is a list of web pages you can refer to:

- https://github.com/sacridini/Awesome-Geospatial
- https://github.com/jerr0328/awesome-geospatial-list

Section 2: Geospatial Analysis Concepts

2

This section represents the main building blocks of this book, where you'll be introduced to Python's role in the industry with different code examples and data editing concepts. You'll learn about geospatial products and how they can be applied to solve problems. Moving on, you'll see how to practically work with remote sensing data using Python. At the end of this section, you'll learn how elevation data can be used in any geospatial format to analyze 3D features.

This section includes the following chapters:

- Chapter 4, *Geospatial Python Toolbox*
- Chapter 5, *Python and Geographic Information Systems*
- Chapter 6, *Python and Remote Sensing*
- Chapter 7, *Python and Elevation Data*

Geospatial Python Toolbox

4

The first three chapters of this book covered the history of geospatial analysis, the types of geospatial data that are used by analysts, and the major software and libraries found within the geospatial industry. We used some simple Python examples here and there to illustrate certain points, but we mainly focused on the field of geospatial analysis, independent of any specific technology. Starting here, we will be using Python to conquer geospatial analysis and we will continue with that approach for the rest of this book. This chapter explains the software you will need in your toolbox to do just about anything you want in the geospatial field.

We'll discover the Python libraries that are used to access the different types of data that were found in the vector data and raster data sections of Chapter 2, *Learning Geospatial Data*. Some of these libraries are pure Python, as well as some of the bindings to the different software packages that we looked at in Chapter 3, *The Geospatial Technology Landscape*.

In this chapter, we will cover the following topics:

- Installing third-party Python modules
- Python virtual environment
- Conda
- Docker
- Python networking libraries for acquiring data
- Python tag-based parsers
- Python JSON libraries
- OGR
- PyShp
- DBFPY
- Shapely
- GDAL
- Fiona

- NumPy
- GeoPandas
- Python Imaging Library (PIL)
- PNGCanvas
- ReportLab
- GeoPDF
- Python NetCDF libraries
- Python HDF libraries
- OSMnx
- Spatial indexing libraries
- Jupyter
- Conda

We will examine pure Python solutions whenever possible. Python is a very capable programming language, but some operations, particularly in remote sensing, are too computationally intensive and therefore are impractical when it comes to using pure Python or other interpreted languages. Fortunately, every aspect of geospatial analysis can be addressed in some way through Python, even if it is binding to a highly efficient C/C++/other compiled-language library.

We will avoid using broad scientific libraries that cover other domains beyond geospatial analysis to keep the solutions as simple as possible. There are many reasons to use Python for geospatial analysis, but one of the strongest arguments is its portability.

Furthermore, Python has been ported to Java as the Jython distribution and to the .NET **Common Language Runtime (CLR)** as IronPython. Python also has versions such as Stackless Python for massively concurrent programs. There are versions of Python that are designed to run on cluster computers for distributed processing. Python is also available on many hosted application servers that do not allow you to install custom executables, such as the Google App Engine platform, which has a Python API.

Technical requirements

- Python 3.6 or higher
- RAM: Minimum 6 GB (Windows), 8 GB (macOS) recommended 8 GB
- Storage: Minimum 7200 RPM SATA with 20 GB of available space; recommended SSD with 40 GB of available space
- Processor: Minimum Intel Core i3 2.5 GHz; recommended Intel Core i5

Installing third-party Python modules

Modules written in pure Python (using the standard library) will mostly run on any of the 20 platforms that the Python (https://www.python.org/) website mentions. Each time you add a third-party module that relies on bindings to external libraries in other languages, you reduce Python's inherent portability. You also add a layer of complexity to fundamentally change the code by adding another language into the mix. Pure Python keeps things simple. Also, Python bindings to external libraries tend to be automatically or semi-automatically generated.

These automatically generated bindings are very generic and esoteric, and they simply connect Python to a C/C++ API using the method names from that API, instead of following the best practices for Python. There are, of course, notable exceptions to this approach that are driven by project requirements which may include speed, unique library features, or frequently updated libraries where an automatically generated interface is preferable.

We'll make a distinction between modules that are included as a part of Python's standard library and modules that must be installed. In Python, the `words` module and library are used interchangeably. To install libraries, you either get them from the **Python Package Index (PyPI)** or in the case of a lot of geospatial modules, you download a specialized installer.

PyPI acts as the official software repository for libraries and offers some easy-to-use setup programs that simplify installing packages. You can use the `easy_install` program, which is especially good on Windows or the `pip` program that's more commonly found on Linux and Unix systems. Once it's installed, you can then install third-party packages by running the following code:

```
easy_install <package name>
```

For installing `pip`, run the following code:

```
pip install <package name>
```

This book will provide links and installation instructions for open source packages that are not available on the PyPI. You can manually install third-party Python modules by downloading the Python source code and putting it in your current working directory, or you can put it in your Python `site-packages` directory. These two directories are available in Python's search path when you try to import a module. If you put a module in your current working directory, it'll only be available when you start Python from that directory.

If you put it in your `site-packages` directory, it'll be available every time you start Python. The `site-packages` directory is specifically meant for third-party modules. To locate the `site-packages` directory for your installation, you need to ask Python's `sys` module. The `sys` module has a `path` attribute that has a list of all the directories in Python's search path. The `site-packages` directory should be the last one. You can locate it by specifying an index of −1, as shown in the following code:

```
>>> import sys
>>> sys.path[-1]
'C:\\Python34\\lib\\site-packages'
```

If that call doesn't return the `site-packages` path, just look at the entire list to locate it, as shown in the following code:

```
>> sys.path
['', 'C:\\WINDOWS\\system32\\python34.zip', 'C:\\Python34\\DLLs',
'C:\\Python34\\lib', 'C:\\Python34\\lib\\plat-win
', 'C:\\Python34\\lib\\lib-tk', 'C:\\Python34',
'C:\\Python34\\lib\\site-packages']
```

 These installation methods will be used in the rest of this book. You can find the latest Python version, the source code for your platform installation, and compilation instructions at `http://python.org/download/`.

The Python `virtualenv` module allows you to easily create an isolated copy of Python for a specific project without affecting your main Python installation or other projects. Using this module, you can have different projects with different versions of the same library. Once you have a working code base, you can keep it isolated from changes to the modules you used or even Python itself. The `virtualenv` module is simple to use and can be used for any example in this book; however, explicit instructions on its use are not included.

 To get started with `virtualenv`, follow this simple guide: `http://docs.python-guide.org/en/latest/dev/virtualenvs/`.

Python virtualenv

Python geospatial analysis requires that we use a variety of modules with many dependencies. These modules often build on each other using specific versions of C or C++ libraries. You often run into version conflicts as you add Python modules to your system. Sometimes, when you upgrade a particular module, it might break your existing Python program due to changes in the API – or maybe you are running both Python 2 and Python 3 to take advantage of libraries written for each version. What you need is a way to safely install new modules without corrupting a working system or code. The solution to that issue is to use Python virtual environments through the `virtualenv` module.

The Python `virtualenv` module creates isolated, individual Python environments for each project so that you can avoid conflicting modules polluting your main Python installation. You can switch a particular environment on and off by activating it or deactivating it. The `virtualenv` module is efficient in that it doesn't actually copy your entire system Python installation each time you create an environment. Let's get started:

1. Installing `virtualenv` is as simple as running the following code:

   ```
   pip install virtualenv
   ```

2. Then, create a directory for your virtual Python environments. Name it whatever you want:

   ```
   mkdir geospatial_projects
   ```

3. Now, you can create your first virtual environment using the following command:

   ```
   virtualenv geospatial_projects/project1
   ```

4. Then, after entering the following command, you can activate the environment:

   ```
   source geospatial_projects/project1/bin/activate
   ```

5. Now, when you run any Python commands in that directory, it will use the isolated virtual environment. When you're done, you can deactivate that environment with the following simple command:

   ```
   deactivate
   ```

This is how you install, activate for use, and deactivate the `virtualenv` module. There's one other environment you should know about, however. We'll examine that next.

Conda

It's also worth mentioning Conda here, which is an open source, cross-platform package management system that can also create and manage environments similar to `virtualenv`. Conda makes it easy to install complex packages, including geospatial ones. It also works with other languages besides Python, including R, Node.js, and Java.

Conda is available here: `https://docs.conda.io/en/latest/`.

Now, let's check out how to install GDAL so that we can start processing geospatial data.

Installing GDAL

The **Geospatial Data Abstraction Library** (**GDAL**), which includes OGR, is critical to many of the examples in this book and is also one of the more complicated Python setups. For these reasons, we'll discuss it separately here. The latest GDAL bindings are available on PyPI; however, the installation requires a few more steps because of additional resources that are needed by the GDAL library.

There are three ways to install GDAL for use with Python. You can use any one of them:

- Compile it from the source code.
- Install it as part of a larger software package.
- Install a binary distribution and then the Python bindings.

If you have experience with compiling C libraries as well as the required compiler software, then the first option gives you the most control. However, it is not recommended if you just want to get going as quickly as possible, because even experienced software developers can find compiling GDAL and the associated Python bindings challenging. Instructions for compiling GDAL on leading platforms can be found at `http://trac.osgeo.org/gdal/wiki/BuildHints`. There are also basic build instructions on the PyPI GDAL page; have a look at `https://pypi.python.org/pypi/GDAL`.

The second option is by far the quickest and easiest. The **Open Source Geospatial Foundation** (**OSGeo**) distributes an installer called OSGeo4W, which installs all of the top open source geospatial packages on Windows at the click of a button. OSGeo4W can be found at `http://trac.osgeo.org/osgeo4w/`.

While these packages are the easiest to work with, they come with their own version of Python. If you already have Python installed, then having another Python distribution just to use certain libraries may be problematic. In that case, the third option may be for you.

The third option installs a pre-compiled binary specific to your Python version. This method is the best compromise between ease of installation and customization. The catch is that you must make sure the binary distributions and the corresponding Python bindings are compatible with each other, your Python version, and in many cases your operating system configuration.

Windows

The installation of GDAL for Python on Windows becomes easier and easier each year. To install GDAL on Windows, you must check whether you are running the 32-bit or 64-bit version of Python:

1. To do so, just start your Python interpreter at a Command Prompt, as shown in the following code:

```
Python 3.4.2 (v3.4.2:ab2c023a9432, Oct 6 2014, 22:15:05) [MSC
v.1600
32 bit (Intel)] on win32
Type "help", "copyright", "credits" or "license" for more
information.
```

2. Based on this instance, we can see that Python is version 3.4.2 for `win32`, which means it is the 32-bit version. Once you have this information, go to the following URL: `http://www.lfd.uci.edu/~gohlke/pythonlibs/#gdal`.

3. This web page contains Python Windows binaries and bindings for nearly every open source scientific library. On that web page, in the GDAL section, find the release that matches your version of Python. The release names use the abbreviation `cp` for C Python, followed by the major Python version number and either `win32` for 32-bit Windows or `win_amd64` for 64-bit Windows.

> In the previous example, we would download the file named
> `GDAL-1.11.3-cp34-none-win32.whl`.

4. This download package is in the newer Python `pip` wheel format. To install it, simply open a Command Prompt and type in the following code:

```
pip install GDAL-1.11.3-cp34-none-win32.whl
```

5. Once the package has been installed, open your Python interpreter and run the following commands to verify that GDAL is installed by checking its version:

```
Python 3.4.2 (v3.4.2:ab2c023a9432, Oct 6 2014, 22:15:05) [MSC
v.1600 32 bit (Intel)] on win32
Type "help", "copyright", "credits" or "license" for more
information.
>>> from osgeo import gdal
>>> gdal.__version__
1.11.3
```

Now, GDAL should return its version as 1.11.3.

> If you have trouble installing modules using easy_install or pip and PyPI, try to download and install the wheel package from the same site as the GDAL example.

Linux

GDAL installation on Linux varies widely by distribution. The following https://gdal.org binaries web page lists the installation instructions for several distributions: http://trac.osgeo.org/gdal/wiki/DownloadingGdalBinaries. Let's get started:

1. Typically, your package manager will install both GDAL and Python bindings. For example, on Ubuntu, to install GDAL, you need to run the following code:

```
sudo apt-get install gdal-bin
```

2. Then, to install the Python bindings, you can run the following command:

```
sudo apt-get install python3-gdal
```

3. Most Linux distributions are set up to compile software already, and their instructions are much simpler than those on Windows.

4. Depending on the installation, you may have to import gdal and ogr as part of the osgeo package, as shown in the following command:

```
>>> from osgeo import gdal
>>> from osgeo import ogr
```

macOS X

To install GDAL on macOS X, you can also use the Homebrew package management system, which is available at `http://brew.sh/`.

Alternatively, you can use the MacPorts package management system, which is available at `https://www.macports.org/`.

Both of these systems are well-documented and contain GDAL packages for Python 3. You only really need them for libraries that require a properly compiled binary written in C that has a lot of dependencies and includes many of the scientific and geospatial libraries.

Python networking libraries for acquiring data

The vast majority of geospatial data sharing is accomplished via the internet, and Python is well equipped when it comes to networking libraries for almost any protocol. Automated data downloads are often an important step in automating a geospatial process. Data is typically retrieved from a website's **Uniform Resource Locator** (**URL**) or **File Transfer Protocol** (**FTP**) server and, because geospatial datasets often contain multiple files, data is often distributed as ZIP files.

A nice feature of Python is its concept of a file-like object. Most Python libraries that read and write data use a standard set of methods that allow you to access data from different types of resources, as if you were writing a simple file on disk. The networking modules in the Python standard library use this convention as well. The benefit of this approach is that it allows you to pass file-like objects to other libraries and methods, which recognize the convention without a lot of setup for different types of data that are distributed in different ways.

The Python urllib module

The Python `urllib` package is designed for simple access to any file with a URL address. The `urllib` package in Python 3 consists of several modules that handle different parts of managing web requests and responses. These modules implement some of Python's file-like object conventions, starting with its `open()` method. When you call `open()`, it prepares a connection to the resource but does not access any data. Sometimes, you just want to grab a file and save it to disk, instead of accessing it in memory. This function is available through the `urllib.request.retrieve()` method.

The following example uses the `urllib.request.retrieve()` method to download the zipped shapefile named `hancock.zip`, which is used in other examples. We define the URL and the local filename as variables. The URL is passed as an argument, as well as the filename we want to use, to save it to our local machine, which, in this case, is just `hancock.zip`:

```
>>> import urllib.request
>>> import urllib.parse
>>> import urllib.error
>>> url = "https://github.com/GeospatialPython/
Learn/raw/master/hancock.zip"
>>> fileName = "hancock.zip"
>>> urllib.request.urlretrieve(url, fileName)
('hancock.zip', <httplib.HTTPMessage instance at 0x00CAD378>)
```

The message from the underlying `httplib` module confirms that the file was downloaded to the current directory. The URL and filename could have been passed to the `retrieve()` method directly as strings as well. If you specify just the filename, the download saves to the current working directory. You can also specify a fully qualified pathname to save it somewhere else. You can also specify a callback function as a third argument, which will receive download status information for the file so that you can create a simple download status indicator or perform some other action.

The `urllib.request.urlopen()` method allows you to access an online resource with more precision and control. As we mentioned previously, it implements most of the Python file-like object methods with the exception of the `seek()` method, which allows you to jump to arbitrary locations within a file. You can read a file online one line at a time, read all the lines as a list, read a specified number of bytes, or iterate through each line of the file. All of these functions are performed in memory, so you don't have to store the data on disk. This ability is useful for accessing frequently updated data online that you may want to process without saving to disk.

In the following example, we demonstrate this concept by accessing the **United States Geological Survey** (**USGS**) earthquake feed to view all of the earthquakes in the world that have occurred within the last hour. This data is distributed as a **Comma-Separated Value** (**CSV**) file, which we can read line by line like a text file. CSV files are similar to spreadsheets and can be opened in a text editor or spreadsheet program:

1. First, you need to open the URL and read the header with the column names in the file.

2. Then, you need to read the first line, which contains a record of a recent earthquake, as shown in the following lines of code:

```
>>> url = "http://earthquake.usgs.gov/earthquakes/feed/v1.0/
summary/all_hour.csv"
>>> earthquakes = urllib.request.urlopen(url)
>>> earthquakes.readline()
'time,latitude,longitude,depth,mag,magType,nst,gap,dmin,rms,net,
id,updated,place
\n'
>>> earthquakes.readline()
'2013-06-14T14:37:57.000Z,64.8405,-147.6478,13.1,0.6,Ml,
6,180,0.09701805,0.2,ak,
ak10739050,2013-06-14T14:39:09.442Z,"3km E of Fairbanks,
Alaska"\n'
```

3. We can also iterate through this file, which is a memory-efficient way to read through large files.

4. If you are running this example in the Python interpreter, you will need to press the *Enter* or *return* key twice to execute the loop. This action is necessary because it signals to the interpreter that you are done building the loop. In the following example, we abbreviate the output:

```
>>> for record in earthquakes: print(record)
2013-06-14T14:30:40.000Z,62.0828,-145.2995,22.5,1.6,
Ml,8,108,0.08174669,0.86,ak,
ak10739046,2013-06-14T14:37:02.318Z,"13km ESE of Glennallen,
Alaska"
...
2013-06-14T13:42:46.300Z,38.8162,-122.8148,3.5,0.6,
Md,,126,0.00898315,0.07,nc,nc
72008115,2013-06-14T13:53:11.592Z,"6km NW of The Geysers,
California"
```

The Python requests module

The `urllib` module has been around for a long time. Another third-party module has been developed to make common HTTP requests even easier. The `requests` module has the following features:

- Keep-alive and connection pooling
- International domains and URLs
- Sessions with cookie persistence
- Browser-style SSL verification

- Automatic content decoding
- Basic/digest authentication
- Elegant key/value cookies
- Automatic decompression
- Unicode response bodies
- HTTP(S) proxy Support
- Multipart file uploads
- Streaming downloads
- Connection timeouts
- Chunked requests
- `.netrc` support

In the following example, we'll download the same ZIP file we downloaded with the `urllib` module, except this time using the `requests` module. First, we need to install the `requests` module:

```
pip install requests
```

Then, we can import it:

```
import requests
```

Then, we can set up our variables for the URL and the output filename:

```
url = "https://github.com/GeospatialPython/Learning/raw/master/hancock.zip"
fileName = "hancock.zip"
```

Retrieving the ZIP file is as simple as using the `requests` module's `get()` method:

```
r = requests.get(url)
```

Now, we can get the content from the `.zip` file and write it to our output file:

```
with open(fileName, 'wb') as f:
    f.write(r.content)
```

The `requests` module has many more advanced features that are just as easy to use as this example. Now that we know how to get information via the HTTP protocol, let's examine the FTP protocol which is often used to access geospatial data from online archives.

FTP

FTP allows you to browse an online directory and download data using FTP client software. Until around 2004, when geospatial web services became very common, FTP was one of the most common ways to distribute geospatial data. FTP is less common now, but you occasionally encounter it when you're searching for data. Once again Python's batteries-included standard library has a reasonable FTP module called `ftplib` with a main class called `FTP()`.

In the following example, we will do the following:

1. We will access an FTP server hosted by the US **National Oceanic and Atmospheric Administration (NOAA)** to access a text file containing data from the **Deep-ocean Assessment and Reporting of Tsunamis (DART)** buoy network that's used to watch for tsunamis around the world. This particular buoy is off the coast of Peru.

2. We'll define the server and the directory path, and then we will access the server. All FTP servers require a username and password. Most public servers have a user called anonymous with the password as anonymous, just like this one does.

3. Using Python's `ftplib`, you can just call the `login()` method without any arguments to log in as the default anonymous user. Otherwise, you can add the username and password as string arguments.

4. Once we're logged in, we'll change to the directory containing the DART datafile.

5. To download the file, we'll open up a local file called out and pass its `write()` method as a callback function to the `ftplib.ftp.retrbinary()` method, which simultaneously downloads the file and writes it to our local file.

6. Once the file has been downloaded, we can close it to save it.

7. Then, we'll read the file and look for the line containing the latitude and longitude of the buoy to make sure that the data was downloaded successfully, as shown in the following lines of code:

```
import ftplib

server = "ftp.ngdc.noaa.gov"
dir = "hazards/DART/20070815_peru"
fileName = "21415_from_20070727_08_55_15_tides.txt"
ftp = ftplib.FTP(server)
ftp.login()
ftp.cwd(dir)

with open(fileName, "wb") as out:
    ftp.retrbinary("RETR " + fileName, out.write)
```

```
with open(fileName) as dart:
  for line in dart:
      if "LAT, " in line:
          print(line)
          break
```

The output is:

LAT,   LON   50.1663   171.8360

In this example, we opened the local file in binary write mode, and we used the retrbinary() ftplib method, as opposed to retrlines(), which uses ASCII mode. The binary mode works for both ASCII and binary files, so it's always a safer bet. In fact, in Python, the binary read and write modes for a file are only required on Windows.

If you are just downloading a simple file from an FTP server, many FTP servers have a web interface as well. In that case, you could use urllib to read the file. FTP URLs use the following format to access data:

```
ftp://username:password@server/directory/file
```

This format is insecure for password-protected directories because you are transmitting your login information over the internet. But for anonymous FTP servers, there is no additional security risk. To demonstrate this, the following example accesses the same file that we just saw but by using urllib instead of ftplib:

```
>>> dart = urllib.request.urlopen("ftp://" + server + "/" + dir +
"/" + fileName)
>>> for line in dart:
... line = str(line, encoding="utf8")
... if "LAT," in line:
... print(line)
... break
...
LAT, LON 50.1663 171.8360
```

Now that we can download files, let's learn how to decompress them.

ZIP and TAR files

Geospatial datasets often consist of multiple files. For this reason, they are often distributed as ZIP or TAR file archives. These formats can also compress data, but their ability to bundle multiple files is the primary reason they are used for geospatial data. While the TAR format doesn't contain a compression algorithm, it incorporates gzip compression and offers it as a program option. Python has standard modules for reading and writing both ZIP and TAR archives. These modules are called `zipfile` and `tarfile`, respectively.

The following example extracts the `hancock.shp`, `hancock.shx`, and `hancock.dbf` files contained in the `hancock.zip` file we downloaded using `urllib` for use in the previous examples. This example assumes that the ZIP file is in the current directory:

```
>>> import zipfile
>>> zip = open("hancock.zip", "rb")
>>> zipShape = zipfile.ZipFile(zip)
>>> shpName, shxName, dbfName = zipShape.namelist()
>>> shpFile = open(shpName, "wb")
>>> shxFile = open(shxName, "wb")
>>> dbfFile = open(dbfName, "wb")
>>> shpFile.write(zipShape.read(shpName))
>>> shxFile.write(zipShape.read(shxName))
>>> dbfFile.write(zipShape.read(dbfName))
>>> shpFile.close()
>>> shxFile.close()
>>> dbfFile.close()
```

This example is more verbose than necessary for clarity. We can shorten this example and make it more robust by using a `for` loop around the `zipfile.namelist()` method without explicitly defining the different files as variables. This method is a more flexible and Pythonic approach, and could be used on ZIP archives with unknown contents, as shown in the following lines of code:

```
>>> import zipfile
>>> zip = open("hancock.zip", "rb")
>>> zipShape = zipfile.ZipFile(zip)
>>> for fileName in zipShape.namelist():
...     out = open(fileName, "wb")
...     out.write(zipShape.read(fileName))
...     out.close()
>>>
```

Now that you understand the basics of the `zipfile` module, let's take the files we just unzipped and create a TAR archive with them. In this example, when we open the TAR archive for writing, we specify the write mode as `w:gz` for gzipped compression. We also specify the file extension as `tar.gz` to reflect this mode, as shown in the following lines of code:

```
>>> import tarfile
>>> tar = tarfile.open("hancock.tar.gz", "w:gz")
>>> tar.add("hancock.shp")
>>> tar.add("hancock.shx")
>>> tar.add("hancock.dbf")
>>> tar.close()
```

We can extract the files using the simple `tarfile.extractall()` method. First, we open the file using the `tarfile.open()` method and then extract it, as shown in the following lines of code:

```
>>> tar = tarfile.open("hancock.tar.gz", "r:gz")
>>> tar.extractall()
>>> tar.close()
```

We'll work on one more example by combining elements we've learned in this chapter as well as the elements in the vector data section of Chapter 2, *Learning Geospatial Data*. We'll read the bounding box coordinates from the `hancock.zip` file without ever saving it to disk. We'll use the power of Python's file-like object convention to pass around the data. Then, we'll use Python's `struct` module to read the bounding box, like we did in Chapter 2, *Learning Geospatial Data*.

In this case, we read the unzipped `.shp` file into a variable and access the data using Python array slicing by specifying the starting and ending indexes of the data separated by a colon (`:`). We are able to use list slicing because Python allows you to treat strings as lists. In this example, we also use Python's `StringIO` module to temporarily store data in memory in a file-like object that implements various methods, including the `seek()` method, which is absent from most Python networking modules, as shown in the following lines of code:

```
>>> import urllib.request
>>> import urllib.parse
>>> import urllib.error
>>> import zipfile
>>> import io
>>> import struct
>>> url =
"https://github.com/GeospatialPython/Learn/raw/master/hancock.zip"
>>> cloudshape = urllib.request.urlopen(url)
```

```
>>> memoryshape = io.BytesIO(cloudshape.read())
>>> zipshape = zipfile.ZipFile(memoryshape)
>>> cloudshp = zipshape.read("hancock.shp")
# Access Python string as an array
>>> struct.unpack("<dddd", cloudshp[36:68])
(-89.6904544701547, 30.173943486533133, -89.32227546981174,
30.6483914869749)
```

As you can see from the examples so far, Python's standard library packs a lot of punch. Most of the time, you don't have to download a third-party library just to access a file online.

Python markup and tag-based parsers

Tag-based data, particularly different XML dialects, have become a very popular way to distribute geospatial data. Formats that are both machine and human-readable are generally easy to work with, though they sacrifice storage efficiency for usability. These formats can become unmanageable for very large datasets but work very well in most cases.

While most formats are some form of XML (such as KML or GML), there is a notable exception. The **Well-Known Text** (**WKT**) format is fairly common but uses external markers and square brackets ([ ]) to surround data instead of tags in angled brackets around data like XML does.

Python has standard library support for XML, as well as some excellent third-party libraries available. Proper XML formats all follow the same structure, so you can use a generic XML library to read it. Because XML is text-based, it is often easy to write it as a string instead of using an XML library. The vast majority of applications that output XML do so in this way.

The primary advantage of using XML libraries for writing XML is that your output is usually validated. It is very easy to create an error while creating your own XML format. A single missing quotation mark can derail an XML parser and throw an error for somebody trying to read your data. When these errors happen, they virtually render your dataset useless. You will find that this problem is very common among XML-based geospatial data. What you'll discover is that some parsers are more forgiving with incorrect XML than others. Often, reliability is more important than speed or memory efficiency.

The analysis that's available at http://lxml.de/performance.html provides benchmarks for memory and speed among the different Python XML parsers.

The minidom module

The Python `minidom` module is a very old and simple to use XML parser. It is part of Python's built-in set of XML tools in the XML package. It can parse XML files or XML that's been fed in as a string. The `minidom` module is best for small to medium-sized XML documents of less than about 20 MB before speed begins to decrease.

To demonstrate the `minidom` module, we'll use a sample KML file, which is a part of Google's KML documentation that you can download. The data that's available at the following link represents time-stamped point locations that have been transferred from a GPS device: `https://github.com/GeospatialPython/Learn/raw/master/time-stamp-point.kml`. Let's get started:

1. First, we'll parse this data by reading it in from the file and creating a `minidom` parser object. The file contains a series of `<Placemark>` tags, which contain a point and a timestamp at which that point was collected. So, we'll get a list of all of the `Placemarks` in the file, and we can count them by checking the length of that list, as shown in the following lines of code:

   ```
   >>> from xml.dom import minidom
   >>> kml = minidom.parse("time-stamp-point.kml")
   >>> Placemarks = kml.getElementsByTagName("Placemark")
   >>> len(Placemarks)
   361
   ```

2. As you can see, we retrieved all `Placemarks`, which totaled `361`. Now, let's take a look at the first `Placemark` element in the list:

   ```
   >>> Placemarks[0]
   <DOM Element: Placemark at 0x2045a30>
   ```

   Each `<Placemark>` tag is now a DOM element data type. To really see what that element is, we call the `toxml()` method, as follows:

   ```
   >>> Placemarks[0].toxml()
   u'<Placemark>\n <TimeStamp>\n \<when>2007-01-14T21:05:02Z</when>\n
   </TimeStamp>\n <styleUrl>#paddle-a</styleUrl>\n <Point>\n
   <coordinates>-122.536226,37.86047,0</coordinates>\n
   </Point>\n </Placemark>'
   ```

3. The `toxml()` function outputs everything contained in the `Placemark` tag as a string object. If we want to print this information to a text file, we can call the `toprettyxml()` method, which would add additional indentation to make the XML more readable.

4. Now, what if we want to grab just the coordinates from this placemark? The coordinates are buried inside the `coordinates` tag, which is contained in the `point` tag and nested inside the `Placemark` tag. Each element of a `minidom` object is called a **node**. Nested nodes are called children or child nodes. The child nodes include more than just tags – they can also include whitespace separating tags, as well as the data inside the tags. So, we can drill down to the `coordinates` tag using the tag name, but then we'll need to access the `data` node. All the `minidom` elements have `childNodeslist`, as well as a `firstChild()` method to access the first node.

5. We'll combine these methods to get to the `data` attribute of the first coordinate's `data` node, which we reference using index 0 in the list of coordinate stags:

```
>>> coordinates =
Placemarks[0].getElementsByTagName("coordinates")
>>> point = coordinates[0].firstChild.data
>>> point
u'-122.536226,37.86047,0'
```

If you're new to Python, you'll notice that the text output in these examples is tagged with the letter u. This markup is how Python denotes Unicode strings that support internationalization to multiple languages with different character sets. Python 3.4.3 changes this convention slightly and leaves Unicode strings unmarked while marking UTF-8 strings with a b.

6. We can go a little further and convert this `point` string into usable data by splitting the string and converting the resulting strings into Python float types, as shown here:

```
>>> x,y,z = point.split(",")
>>> x
u'-122.536226'
>>> y
u'37.86047'
>>> z
u'0'
>>> x = float(x)
>>> y = float(y)
>>> z = float(z)
>>> x,y,z
(-122.536226, 37.86047, 0.0)
```

7. Using Python list comprehension, we can perform this operation in a single step, as you can see in the following lines of code:

```
>>> x,y,z = [float(c) for c in point.split(",")]
>>> x,y,z
(-122.536226, 37.86047, 0.0)
```

This example scratches the surface of what the `minidom` library can do. For a great tutorial on this library, have a look at the following tutorial: `https://www.edureka.co/blog/python-xml-parser-tutorial/`.

ElementTree

The `minidom` module is pure Python, easy to work with, and has been around since Python 2.0. However, Python 2.5 added a more efficient yet high-level XML parser to the standard library called `ElementTree`. `ElementTree` is interesting because it has been implemented in multiple versions.

There is a pure Python version and a faster version written in C called `cElementTree`. You should use `cElementTree` wherever possible, but it's possible that you may be on a platform that doesn't include the C-based version. When you import `cElementTree`, you can test to see if it's available and fall back to the pure Python version if necessary:

```
try:
    import xml.etree.cElementTree as ET
except ImportError:
    import xml.etree.ElementTree as ET
```

One of the great features of `ElementTree` is its implementation of a subset of the XPath query language. XPath is short for XML Path and allows you to search an XML document using a path-style syntax. If you work with XML frequently, learning XPath is essential. You can learn more about XPath at the following link: `https://www.w3schools.com/xml/xpath_intro.asp`.

One catch with this feature is that if the document specifies a namespace, as most XML documents do, you must insert that namespace into queries. `ElementTree` does not automatically handle the namespace for you. Your options are to manually specify it or try to extract it using string parsing from the root element's tag name.

We'll repeat the `minidomXML` parsing example using `ElementTree`:

1. First, we'll parse the document and then we'll manually define the KML namespace; later, we'll use an XPath expression and the `find()` method to find the first `Placemark` element.
2. Finally, we'll find the coordinates and the child node and then grab the text containing the latitude and longitude.

In both cases, we could have searched directly for the `coordinates` tag. But, by grabbing the `Placemark` element, it gives us the option of grabbing the corresponding timestamp child element later, if we so choose, as shown in the following lines of code:

```
>>> tree = ET.ElementTree(file="time-stamp-point.kml")
>>> ns = "{http://www.opengis.net/kml/2.2}"
>>> placemark = tree.find(".//%sPlacemark" % ns)
>>> coordinates =
placemark.find("./{}Point/{}coordinates".format(ns, ns))
>>> coordinates.text
'-122.536226,37.86047,0'
```

In this example, notice that we used the Python string formatting syntax, which is based on the string formatting concept found in C. When we defined the XPath expression for the placemark variable, we used the `%` placeholder to specify the insertion of a string. Then, after the string, we used the `%` operator followed by a variable name to insert the `ns` namespace variable where the placeholder is. In the `coordinates` variable, we used the `ns` variable twice, so we specified a tuple containing `ns` twice after the string.

> String formatting is a simple yet extremely powerful and useful tool in Python that's worth learning. You can find more information in Python's documentation online at the following link: `https://docs.python.org/3.4/library/string.html`.

Building XML using ElementTree and Minidom

Most of the time, XML can be built by concatenating strings, as you can see in the following command:

```
xml = "<?xml version="1.0" encoding="utf-8"?>"
xml += "<kml xmlns="http://www.opengis.net/kml/2.2">"
xml += " <Placemark>"
xml += " <name>Office</name>"
xml += " <description>Office Building</description>"
xml += " <Point>"
```

```
xml += " <coordinates>"
xml += " -122.087461,37.422069"
xml += " </coordinates>"
xml += " </Point>"
xml += " </Placemark>"
xml += "</kml>"
```

However, this method can be quite prone to typos, which creates invalid XML documents. A safer way is to use an XML library. Let's build this simple KML document using ElementTree:

1. We'll define the root KML element and assign it a namespace.

2. Then, we'll systematically append sub elements to the root, wrap the elements as an ElementTree object, declare the XML encoding, and write it out to a file called placemark.xml, as shown in the following lines of code:

```
>>> root = ET.Element("kml")
>>> root.attrib["xmlns"] = "http://www.opengis.net/kml/2.2"
>>> placemark = ET.SubElement(root, "Placemark")
>>> office = ET.SubElement(placemark, "name")
>>> office.text = "Office"
>>> point = ET.SubElement(placemark, "Point")
>>> coordinates = ET.SubElement(point, "coordinates")
>>> coordinates.text = "-122.087461,37.422069, 37.422069"
>>> tree = ET.ElementTree(root)
>>> tree.write("placemark.kml",
xml_declaration=True,encoding='utf-8',method="xml")
```

The output is identical to the previous string building example, except that ElementTree does not indent the tags but rather writes it as one long string. The minidom module has a similar interface, which is documented in the book *Dive Into Python*, by Mark Pilgrim, which was referenced in the minidom example that we just saw.

XML parsers such as minidom and ElementTree work very well on perfectly formatted XML documents. Unfortunately, the vast majority of XML documents out there don't follow these rules and contain formatting errors or invalid characters. You'll find that you are often forced to work with this data and must resort to extraordinary string parsing techniques to get the small subset of data you actually need. But thanks to Python and Beautiful Soup, you can elegantly work with bad and even terrible tag-based data.

Beautiful Soup is a module that was specifically designed to robustly handle broken XML. It is oriented toward HTML, which is notorious for incorrect formatting but works with other XML dialects too. Beautiful Soup is available on PyPI, so use either `easy_install` or `pip` to install it, as you can see in the following command:

```
easy_install beautifulsoup4
```

Alternatively, you can execute the following command:

```
pip install beautifulsoup4
```

Then, to use it, you simply import it:

```
>>> from bs4 import BeautifulSoup
```

To try it out, we'll use a **GPS Exchange Format** (**GPX**) tracking file from a smartphone application, which has a glitch and exports slightly broken data. You can download this sample file from `https://raw.githubusercontent.com/GeospatialPython/Learn/master/broken_data.gpx`.

This 2,347-line data file is in pristine condition except that it is missing a closing `</trkseg>` tag, which should be located at the very end of the file, just before the closing `</trk>` tag. This error was caused by a data export function in the source program. This defect is most likely a result of the original developer manually generating the GPX XML on export and forgetting the line of code that adds this closing tag. Watch what happens if we try to parse this file with `minidom`:

```
>>> gpx = minidom.parse("broken_data.gpx")
Traceback (most recent call last):
File "<stdin>", line 1, in <module>
File "C:\Python34\lib\xml\dom\minidom.py", line 1914, in parse
return expatbuilder.parse(file)
File "C:\Python34\lib\xml\dom\expatbuilder.py", line 924, in
parse
result = builder.parseFile(fp)
File "C:\Python34\lib\xml\dom\expatbuilder.py", line 207, in
parseFile
parser.Parse(buffer, 0)
xml.parsers.expat.ExpatError: mismatched tag: line 2346, column 2
```

As you can see from the last line in the error message, the underlying XML parser in `minidom` knows exactly what the problem is – a `mismatched` tag right at the end of the file. However, it refused to do anything more than report the error. You must have perfectly formed XML or none at all to avoid this.

Now, let's try the more sophisticated and efficient `ElementTree` module with the same data:

```
>>> ET.ElementTree(file="broken_data.gpx")
Traceback (most recent call last):
File "<stdin>", line 1, in <module>
File "C:\Python34\lib\xml\etree\ElementTree.py", line 611, in
__init__
self.parse(file)
File "C:\Python34\lib\xml\etree\ElementTree.py", line 653, in
parse
parser.feed(data)
File "C:\Python34\lib\xml\etree\ElementTree.py", line 1624, in
feed
self._raiseerror(v)
File "C:\Python34\lib\xml\etree\ElementTree.py", line 1488, in
_raiseerror
raise err
xml.etree.ElementTree.ParseError: mismatched tag: line 2346,
column 2
```

As you can see, different parsers face the same problem. Poorly formed XML is an all too common reality in geospatial analysis, and every XML parser assumes that all the XML in the world is perfect, except for one. Enter Beautiful Soup. This library shreds bad XML into usable data without a second thought, and it can handle far worse defects than missing tags. It will work despite missing punctuation or other syntax and will give you the best data it can. It was originally developed for parsing HTML, which is notoriously horrible for being poorly formed, but it works fairly well with XML as well, as shown here:

```
>>> from bs4 import BeautifulSoup
>>> gpx = open("broken_data.gpx")
>>> soup = BeautifulSoup(gpx.read(), features="xml")
>>>
```

No complaints from Beautiful Soup! Just to make sure the data is actually usable, let's try and access some of the data. One of the fantastic features of Beautiful Soup is that it turns tags into attributes of the parse tree. If there are multiple tags with the same name, it grabs the first one. Our sample data file has hundreds of `<trkpt>` tags. Let's access the first one:

```
>>> soup.trkpt
<trkpt lat="30.307267000" lon="-89.332444000">
<ele>10.7</ele><time>2013-05-16T04:39:46Z</time></trkpt>
```

We're now certain that the data has been parsed correctly and that we can access it. If we want to access all of the `<trkpt>` tags, we can use the `findAll()` method to grab them and then use the built-in Python `len()` function to count them, as shown here:

```
>>> tracks = soup.findAll("trkpt")
>>> len(tracks)
2321
```

If we write the parsed data back out to a file, Beautiful Soup outputs the corrected version. We'll save the fixed data as a new GPX file using Beautiful Soup module's `prettify()` method to format the XML with nice indentation, as you can see in the following lines of code:

```
>>> fixed = open("fixed_data.gpx", "w")
>>> fixed.write(soup.prettify())
>>> fixed.close()
```

Beautiful Soup is a very rich library with many more features. To explore it further, visit the Beautiful Soup documentation online at `http://www.crummy.com/software/ BeautifulSoup/bs4/documentation.html`.

While `minidom`, `ElementTree`, and `cElementTree` come with the Python standard library, there is an even more powerful and popular XML library for Python called `lxml`. The `lxml` module provides a Pythonic interface to the `libxml2` and `libxslt` C libraries using the `ElementTree` API. An even better fact is that `lxml` also works with Beautiful Soup to parse bad tag-based data. On some installations, `beautifulsoup4` may require `lxml`. The `lxml` module is available via PyPI but requires some additional steps for the C libraries. More information is available on the `lxml` home page at the following link: `http://lxml.de/`.

Well-Known Text (WKT)

The WKT format has been around for years and is a simple text-based format for representing geometries and spatial reference systems. It is primarily used as a data exchange format by systems that implement the OGC Simple Features for SQL specification. Take a look at the following sample WKT representation of a polygon:

```
POLYGON((0 0,4 0,4 4,0 4,0 0),(1 1, 2 1, 2 2, 1 2,1 1))
```

Currently, the best way to read and write WKT is by using the Shapely library. Shapely provides a very Python-oriented or Pythonic interface to the **Geometry Engine - Open Source (GEOS)** library we described in Chapter 3, *The Geospatial Technology Landscape*.

You can install Shapely using either easy_install or pip. You can also use the wheel from the site we mentioned in the previous section. Shapely has a WKT module which can load and export this data. Let's use Shapely to load the previous polygon sample and then verify that it has been loaded as a polygon object by calculating its area:

```
>>> import shapely.wkt
>>> wktPoly = "POLYGON((0 0,4 0,4 4,0 4,0 0),(1 1, 2 1, 2 2, 1 2,
1 1))"
>>> poly = shapely.wkt.loads(wktPoly)
>>> poly.area
15.0
```

We can convert any Shapely geometry back into a WKT by simply calling its wkt attribute, as shown here:

```
>>> poly.wkt
'POLYGON ((0.0 0.0, 4.0 0.0, 4.0 4.0, 0.0 4.0, 0.0 0.0), (1.0 1.0,
2.0 1.0, 2.0 2.0, 1.0 2.0, 1.0 1.0))'
```

Shapely can also handle the WKT binary counterpart called **Well-Known Binary** (**WKB**), which is used to store WKT strings as binary objects in databases. Shapely loads WKB using its wkb module in the same way as the wkt module, and it can convert geometries by calling that object's wkb attribute.

Shapely is the most Pythonic way to work with WKT data, but you can also use the Python bindings to the OGR library, which we installed earlier in this chapter.

For this example, we'll use a shapefile with one simple polygon, which can be downloaded as a ZIP file. It is available at the following link: https://github.com/GeospatialPython/Learn/raw/master/polygon.zip.

In the following example, we'll open the polygon.shp file from the shapefile dataset, call the required GetLayer() method, get the first (and only) feature, and then export it to WKT:

```
>>> from osgeo import ogr
>>> shape = ogr.Open("polygon.shp")
>>> layer = shape.GetLayer()
>>> feature = layer.GetNextFeature()
>>> geom = feature.GetGeometryRef()
>>> wkt = geom.ExportToWkt()
>>> wkt
```

```
' POLYGON ((-99.904679362176353 51.698147686745074,
-75.010398603076666 46.56036851832075,-75.010398603076666
46.56036851832075,-75.010398603076666 46.56036851832075,
-76.975736557742451 23.246272688996914,-76.975736557742451
23.246272688996914,-76.975736557742451 23.246272688996914,
-114.31715769639194 26.220870210283724,-114.31715769639194
26.220870210283724,-99.904679362176353 51.698147686745074))'
```

Note that with OGR, you would have to read access each feature and export it individually, since the `ExporttoWkt()` method is at the feature level. We can now turn around and read a WKT string using the `wkt` variable containing the export. We'll import it back into `ogr` and get the bounding box, also known as an envelope, of the polygon, as you can see here:

```
>>> poly = ogr.CreateGeometryFromWkt(wkt)
>>> poly.GetEnvelope()
(-114.31715769639194, -75.01039860307667, 23.246272688996914,
51.698147686745074)
```

Shapely and OGR are used for reading and writing valid WKT strings. Of course, just like XML, which is also text, you could manipulate small amounts of WKT as strings in a pinch. Next, we'll look at a modern text format that is becoming very common in the geospatial world.

Python JSON libraries

JavaScript Object Notation (JSON) is rapidly becoming the number one data exchange format across a lot of fields. The lightweight syntax and its similarity to existing data structures in both the JavaScript that Python borrows some data structures from, as well as JavaScript itself, make it a perfect match for Python.

The following GeoJSON sample document contains a single point:

```
{
    "type": "Feature",
    "id": "OpenLayers.Feature.Vector_314",
    "properties": {},
    "geometry": {
        "type": "Point",
        "coordinates": [
            97.03125,
            39.7265625
        ]
    },
    "crs": {
        "type": "name",
```

```
        "properties": {
            "name": "urn:ogc:def:crs:OGC:1.3:CRS84"
        }
    }
}
```

This sample is just a simple point with new attributes, which would be stored in the properties data structure of the geometry. In the preceding example, the ID, coordinates, and CRS information would change depending on your particular dataset.

Let's modify this sample GeoJSON document using Python. First, we'll compact the sample document into a single string to make it easier to handle:

```
>>> jsdata = """{
    "type": "Feature",
    "id": "OpenLayers.Feature.Vector_314",
    "properties": {},
    "geometry": {
        "type": "Point",
        "coordinates": [
            97.03125,
            39.7265625
        ]
    },
    "crs": {
        "type": "name",
        "properties": {
            "name": "urn:ogc:def:crs:OGC:1.3:CRS84"
        }
    }
}"""
```

Now, we can use the GeoJSON `jsdata` string variable we created in the preceding code, in the following examples.

The json module

GeoJSON looks very similar to a nested set of Python's dictionaries and lists. Just for fun, let's just try and use Python's `eval()` function to parse it as Python code:

```
>>> point = eval(jsdata)
>>> point["geometry"]
{'type': 'Point', 'coordinates': [97.03125, 39.7265625]}
```

Wow! That worked! We turned that random GeoJSON string into native Python data in one easy step. Keep in mind that the JSON data format is based on JavaScript syntax, which happens to be similar to Python. Also, as you get deeper into GeoJSON data and work with larger data, you'll find that JSON allows characters that Python does not. Using Python's `eval()` function is considered very insecure as well. But as far as keeping things simple is concerned, note that it doesn't get any simpler than that!

Thanks to Python's drive toward simplicity, the more advanced method doesn't get much more complicated. Let's use Python's `json` module, which is part of the standard library, to turn the same string into Python the right way:

```
>>> import json
>>> json.loads(jsdata)
{u'geometry': {u'type': u'Point', u'coordinates': [97.03125,
39.7265625]}, u'crs': {u'type': u'name', u'properties': {u'name':
u'urn:ogc:def:crs:OGC:1.3:CRS84'}}, u'type': u'Feature', u'id':
u'OpenLayers.Feature.Vector_314',
u'properties':
{}}
```

As a side note, in the previous example, the CRS84 property is a synonym for the common WGS84 coordinate system. The `json` module adds some nice features such as safer parsing and conversion of strings into Unicode. We can export Python data structures to JSON in almost the same way:

```
>>> pydata = json.loads(jsdata)
>>> json.dumps(pydata)
'{"geometry": {"type": "Point", "coordinates": [97.03125,
39.7265625]}, "crs": {"type": "name", "properties": {"name":
"urn:ogc:def:crs:OGC:1.3:CRS84"}}, "type" : "Feature", "id":
"OpenLayers.Feature.Vector_314", "properties":
{}}'
```

When you dump data, it comes out as one long string that's difficult to read. There's a way we can print the data so it is easier to read: by passing the `dumps()` method an indent value:

```
print(json.dumps(pydata, indent=4))

{
    "type": "Feature",
    "id":
    "OpenLayers.Feature.Vector_314",
    "properties": {},
    "geometry": {
        "type": "Point",
```

```
        "coordinates": [
            97.03125,
            39.7265625
        ]
    },
    "crs": {
        "type": "name",
        "properties": {
            "name": "urn:ogc:def:crs:OGC:1.3:CRS84"
        }
    }
}
```

Now that we understand `json` module, let's look at the geospatial version called `geojson`.

The geojson module

We could happily go on reading and writing GeoJSON data using the `json` module forever, but there's an even better way. The `geojson` module that's available on PyPI offers some distinct advantages. For starters, it knows the requirements of the GeoJSON specification, which can save a lot of typing. Let's create a simple point using this module and export it to GeoJSON:

```
>>> import geojson
>>> p = geojson.Point([-92, 37])
```

This time, when we dump the JSON data for viewing, we'll add an indent argument with a value of 4 so that we get nicely indented JSON data that's easier to read:

```
>>> geojs = geojson.dumps(p, indent=4)
>>> geojs
```

Our output is as follows:

```
{
    "type": "Point",
    "coordinates": [
        -92,
        37
    ]
}
POINT (-92 37)
```

Notice that the `geojson` module has an interface for different data types and saves us from setting the type and coordinates attributes manually. Now, imagine if you had a geographic object with hundreds of features. You could programmatically build this data structure instead of building a very large string.

The `geojson` module is also the reference implementation for the Python `geo_interface` convention. This interface allows cooperating programs to exchange data seamlessly and in a Pythonic way without the programmer explicitly exporting and importing GeoJSON strings. So, if we wanted to feed the point we created with the `geojson` module to the Shapely module, we could perform the following command, which reads the `geojson` module's point object straight into Shapely, after which we'll export it as WKT:

```
>>> from shapely.geometry import asShape
>>> point = asShape(p)
>>> point.wkt
'POINT (-92.0000000000000000 37.0000000000000000)'
```

More and more geospatial Python libraries are implementing both the `geojson` and `geo_interface` functionality, including PyShp, Fiona, Karta, and ArcGIS. Third-party implementations exist for QGIS.

GeoJSON is a simple text format that is human and computer-readable. Now, we'll look at some binary vector formats.

OGR

We touched on OGR as a way to handle WKT strings, but its real power is as a universal vector library. This book strives for pure Python solutions, but no single library even comes close to the variety of formats that OGR can process.

Let's read a sample point shapefile using the OGR Python API. The sample shapefile can be downloaded as a ZIP file here:
`https://github.com/GeospatialPython/Learn/raw/master/point.zip`.

This point shapefile has five points with single digit, positive coordinates. The attributes list the order in which the points were created, making it useful for testing. This simple example will read in the point shapefile and loop through each feature; then, it will print the *x* and *y* values of each point, plus the value of the first attribute field:

```
>>> import ogr
>>> shp = ogr.Open("point.shp")
>>> layer = shp.GetLayer()
>>> for feature in layer:
```

```
... geometry = feature.GetGeometryRef()
... print(geometry.GetX(), geometry.GetY(),
feature.GetField("FIRST_FLD"))
...
1.0 1.0 First
3.0 1.0 Second
4.0 3.0 Third
2.0 2.0 Fourth
0.0 0.0 Appended
```

This example is simple, but OGR can become quite verbose as your script becomes more complex. Next, we'll look at a simpler way to deal with shapefiles.

PyShp

PyShp is a simple, pure Python library that reads and writes shapefiles. It doesn't perform any geometry operations and only uses Python's standard library. It's contained in a single file that's easy to move around, squeeze onto small embedded platforms, and modify. It is also compatible with Python 3. It also implements __geo_interface__. The PyShp module is available on PyPI.

Let's repeat the previous OGR example with PyShp:

```
>>> import shapefile
>>> shp = shapefile.Reader("point.shp")
>>> for feature in shp.shapeRecords():
... point = feature.shape.points[0]
... rec = feature.record[0]
... print(point[0], point[1], rec)
...
1.0 1.0 First
3.0 1.0 Second
4.0 3.0 Third
2.0 2.0 Fourth
0.0 0.0 Appended
//
```

dbfpy

Both OGR and PyShp read and write the .dbf files because they are part of the shapefile specification. The .dbf files contain the attributes and fields for the shapefiles. However, both libraries have very basic .dbf support. Occasionally, you will need to do some heavy-duty DBF work. The dbfpy3 module is a pure Python module dedicated to working with .dbf files. It is currently hosted on GitHub. You can force easy_install to find the download by specifying the download file:

```
easy_install -f
    https://github.com/GeospatialPython/dbfpy3/archive/master.zip
```

If you are using pip to install packages, use the following command:

```
pip install
    https://github.com/GeospatialPython/dbfpy3/archive/master.zip
```

The following shapefile has over 600 .dbf records representing US Census Bureau tracts, which make it a good sample for trying out dbfpy: https://github.com/GeospatialPython/Learn/raw/master/GIS_CensusTract.zip.

Let's open up the .dbf file of this shapefile and look at the first record:

```
>>> from dbfpy3 import dbf
>>> db = dbf.Dbf("GIS_CensusTract_poly.dbf")
>>> db[0]
GEODB_OID: 4029 (<type 'int'>)
OBJECTID: 4029 (<type 'int'>)
PERMANE0: 61be9239-8f3b-4876-8c4c-0908078bc597 (<type 'str'>)
SOURCE_1: NA (<type 'str'>)
SOURCE_2: 20006 (<type 'str'>)
SOURCE_3: Census Tracts (<type 'str'>)
SOURCE_4: Census Bureau (<type 'str'>)
DATA_SE5: 5 (<type 'str'>)
DISTRIB6: E4 (<type 'str'>)
LOADDATE: 2007-03-13 (<type 'datetime.date'>)
QUALITY: 2 (<type 'str'>)
SCALE: 1 (<type 'str'>)
FCODE: 1734 (<type 'str'>)
STCO_FI7: 22071 (<type 'str'>)
STATE_NAME: 22 (<type 'str'>)
COUNTY_8: 71 (<type 'str'>)
CENSUST9: 22071001734 (<type 'str'>)
POPULAT10: 1760 (<type 'int'>)
AREASQKM: 264.52661934 (<type 'float'>)
GNIS_ID: NA (<type 'str'>)
```

```
POPULAT11: 1665 (<type 'int'>)
DB2GSE_12: 264526619.341 (<type 'float'>)
DB2GSE_13: 87406.406192 (<type 'float'>)
```

The module quickly and easily gives us both the column names and data values together, as opposed to handling them as separate lists, so that they're easier to manage. Now, let's increment the population field contained in POPULAT10 by 1:

```
>>> rec = db[0]
>>> field = rec["POPULAT10"]
>>> rec["POPULAT10"] = field + 1
>>> rec.store()
>>> del rec
>>> db[0]["POPULAT10"]
1761
```

Keep in mind that both OGR and PyShp can do this same procedure, but dbfp3y makes it a little easier if you are only making a lot of changes to the .dbf files.

Shapely

Shapely was mentioned in the **Well-Known Text (WKT)** section for its import and exportability. However, its true purpose is as a generic geometry library. Shapely is a high-level, Pythonic interface to the GEOS library for geometric operations. In fact, Shapely intentionally avoids reading or writing files. It relies completely on data import and export from other modules and maintains focus on geometry manipulation.

Let's do a quick Shapely demonstration in which we'll define a single WKT polygon and then import it into Shapely. Then, we'll measure the area. Our computational geometry will consist of buffering that polygon by a measure of five arbitrary units, which will return a new, bigger polygon for which we'll measure the area:

```
>>> from shapely import wkt, geometry
>>> wktPoly = "POLYGON((0 0,4 0,4 4,0 4,0 0))"
>>> poly = wkt.loads(wktPoly)
>>> poly.area
16.0
>>> buf = poly.buffer(5.0)
>>> buf.area
174.41371226364848
```

We can then perform a difference in the area of the buffer and the original polygon area, as shown here:

```
>>> buf.difference(poly).area
158.413712264
```

If you can't have pure Python, a Pythonic API as clean as Shapely that packs such a punch is certainly the next best thing.

Fiona

The Fiona library provides a simple Python API around the OGR library for data access and nothing more. This approach makes it easy to use and is less verbose than OGR while using Python. Fiona outputs GeoJSON by default. You can find a wheel file for Fiona at `http://www.lfd.uci.edu/~gohlke/pythonlibs/#fiona`.

As an example, we'll use the `GIS_CensusTract_poly.shp` file from the `dbfpy` example we looked at earlier in this chapter.

First, we'll import `fiona` and Python's `pprint` module to format the output. Then, we'll open the shapefile and check its driver type:

```
>>> import fiona
>>> from pprint import pprint
>>> f = fiona.open("GIS_CensusTract_poly.shp")
>>> f.driver
```

ESRI shapefile

Next, we'll check its coordinate reference system and get the data bounding box, as shown here:

```
>>> f.crs
{'init': 'epsg:4269'}
>>> f.bounds
(-89.8744162216216, 30.161122135135138, -89.1383837783784,
30.661213864864862)
```

Now, we'll view the data schema as `geojson` and format it using the `pprint` module, as you can see in the following lines of code:

```
>>> pprint(f.schema)
{'geometry': 'Polygon',
 'properties': {'GEODB_OID': 'float:11',
 'OBJECTID': 'float:11',
 'PERMANE0': 'str:40',
 'SOURCE_1': 'str:40',
 'SOURCE_2': 'str:40',
 'SOURCE_3': 'str:100',
 'SOURCE_4': 'str:130',
 'DATA_SE5': 'str:46',
 'DISTRIB6': 'str:188',
 'LOADDATE': 'date',
 'QUALITY': 'str:35',
 'SCALE': 'str:52',
 'FCODE': 'str:38',
 'STCO_FI7': 'str:5',
 'STATE_NAME': 'str:140',
 'COUNTY_8': 'str:60',
 'CENSUST9': 'str:20',
 'POPULAT10': 'float:11',
 'AREASQKM': 'float:31.15',
 'GNIS_ID': 'str:10',
 'POPULAT11': 'float:11',
 'DB2GSE_12': 'float:31.15',
 'DB2GSE_13': 'float:31.15'}}
```

Next, let's get a count of the number of features:

```
>>> len(f)
45
```

Finally, we'll print one of the records as formatted GeoJSON, as shown here:

```
pprint(f[1])
{'geometry': {'coordinates': [[[(-89.86412366375093,
30.661213864864862), (-89.86418691770497, 30.660764012731285),
(-89.86443391770518, 30.659652012730202),
...
'type': 'MultiPolygon'},
'id': '1',
'properties': {'GEODB_OID': 4360.0,
'OBJECTID': 4360.0,
'PERMANE0': '9a914eef-9249-44cf-a05f-af4b48876c59',
'SOURCE_1': 'NA',
'SOURCE_2': '20006',
...
'DB2GSE_12': 351242560.967882,
'DB2GSE_13': 101775.283967268},
'type': 'Feature'}
```

GDAL

GDAL is the dominant geospatial library for raster data. Its raster capability is so significant that it is a part of virtually every geospatial toolkit in any language, and Python is no exception to this. To see the basics of how GDAL works in Python, download the following sample raster satellite image as a ZIP file and unzip it: https://github.com/ GeospatialPython/Learn/raw/master/SatImage.zip. Let's open this image and see how many bands it has and how many pixels are present along each axis:

```
>>> from osgeo import gdal
>>> raster = gdal.Open("SatImage.tif")
>>> raster.RasterCount
3
>>> raster.RasterXSize
2592
>>> raster.RasterYSize
2693
```

By viewing it in OpenEV, we can see that the following image has three bands, 2,592 columns of pixels, and 2,693 rows of pixels:

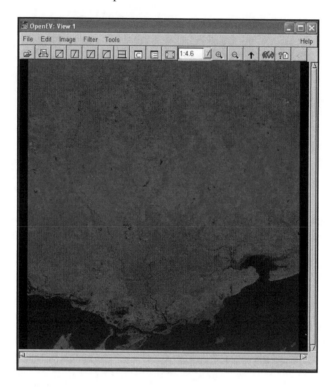

GDAL is an extremely fast geospatial raster reader and writer within Python. It can also reproject images quite well in addition to being able to do a few other tricks. However, the true value of GDAL comes from its interaction with the next Python module, which we'll examine now.

NumPy

NumPy is an extremely fast, multidimensional Python array processor designed specifically for Python and scientific computing but is written in C. It is available via PyPI or as a wheel file (available at http://www.lfd.uci.edu/~gohlke/pythonlibs/#numpy) and can be installed with ease. In addition to its amazing speed, the magic of NumPy includes its interaction with other libraries. NumPy can exchange data with GDAL, Shapely, the **Python Imaging Library (PIL)**, and many other scientific computing Python libraries in other fields.

As a quick example of NumPy's ability, we'll combine it with GDAL to read in our sample satellite image and then create a histogram of it. The interface between GDAL and NumPy is a GDAL module called `gdal_array`, which has NumPy as a dependency. Numeric is the legacy name of the NumPy module. The `gdal_array` module imports NumPy.

In the following example, we'll use `gdal_array`, which imports NumPy, to read the image in as an array, grab the first band, and save it as a JPEG image:

```
>>> from osgeo import gdal_array
>>> srcArray = gdal_array.LoadFile("SatImage.tif")
>>> band1 = srcArray[0]
>>> gdal_array.SaveArray(band1, "band1.jpg", format="JPEG")
```

This operation gives us the following grayscale image in OpenEV:

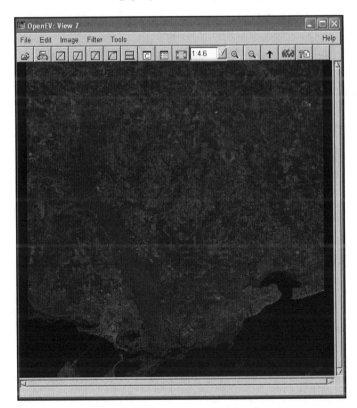

PIL

PIL was originally developed for remote sensing but has evolved as a general image editing library for Python. Like NumPy, it is written in C for speed but is designed specifically for Python. In addition to image creation and processing, it also has a useful raster drawing module. PIL is also available via PyPI; however, in Python 3, you may want to use the Pillow module, which is an upgraded version of PIL. As you'll see in the following example, we can use a Python try statement to import PIL using two possible variations, depending on how you installed it.

In this example, we'll combine PyShp and PIL to rasterize the hancock shapefile from the previous examples and save it as an image. We'll use a world to pixel coordinate transformation similar to our SimpleGIS from Chapter 1, *Learning about Geospatial Analysis with Python*. We'll create an image to use as a canvas in PIL, and then we'll use the PIL ImageDraw module to render the polygon. Finally, we'll save it as a PNG image, as you can see in the following lines of code:

```
>>> try:
>>> import Image
>>> import ImageDraw
>>> except:
>>> from PIL import Image
>>> from PIL import ImageDraw
>>> import shapefile
>>> r = shapefile.Reader("hancock.shp")
>>> xdist = r.bbox[2] - r.bbox[0]
>>> ydist = r.bbox[3] - r.bbox[1]
>>> iwidth = 400
>>> iheight = 600
>>> xratio = iwidth/xdist
>>> yratio = iheight/ydist
>>> pixels = []
>>> for x,y in r.shapes()[0].points:
... px = int(iwidth - ((r.bbox[2] - x) * xratio))
... py = int((r.bbox[3] - y) * yratio)
... pixels.append((px,py))
...
>>> img = Image.new("RGB", (iwidth, iheight), "white")
>>> draw = ImageDraw.Draw(img)
>>> draw.polygon(pixels, outline="rgb(203, 196, 190)",
fill="rgb(198, 204, 189)")
>>> img.save("hancock.png")
```

This example creates the following image:

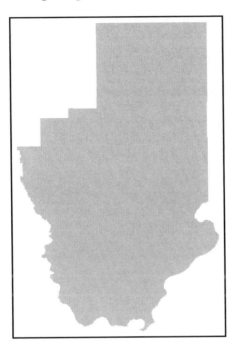

PNGCanvas

Sometimes, you may find that PIL is overkill for your purposes, or you are not allowed to install PIL because you do not have administrative rights to the machine that you're using to install Python modules that have been created and compiled in C. In those cases, you can usually get away with the lightweight pure Python PNGCanvas module. You can install it using `easy_install` or pip.

Using this module, we can repeat the raster shapefile example we performed using PIL but in pure Python, as you can see here:

```
>>> import shapefile
>>> import pngcanvas
>>> r = shapefile.Reader("hancock.shp")
>>> xdist = r.bbox[2] - r.bbox[0]
>>> ydist = r.bbox[3] - r.bbox[1]
>>> iwidth = 400
>>> iheight = 600
>>> xratio = iwidth/xdist
```

```
>>> yratio = iheight/ydist
>>> pixels = []
>>> for x,y in r.shapes()[0].points:
... px = int(iwidth - ((r.bbox[2] - x) * xratio))
... py = int((r.bbox[3] - y) * yratio)
... pixels.append([px,py])
...
>>> c = pngcanvas.PNGCanvas(iwidth,iheight)
>>> c.polyline(pixels)
>>> f = open("hancock_pngcvs.png", "wb")
>>> f.write(c.dump())
>>> f.close()
```

This example gives us a simple outline as PNGCanvas does not have a built-in fill method:

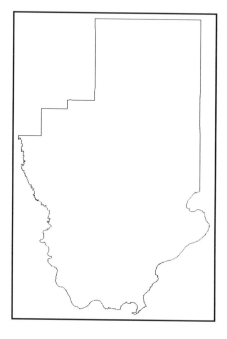

GeoPandas

Pandas is a high-performance Python data analysis library that can handle large datasets that are tabular (similar to a database), ordered/unordered, labeled matrices, or unlabeled statistical data. GeoPandas is simply a geospatial extension to Pandas that builds upon Shapely, Fiona, PyProj, Matplotlib, and Descartes, all of which must be installed. It allows you to easily perform operations in Python, which would otherwise require a spatial database such as PostGIS. You can download a wheel file for GeoPandas from `http://www.lfd.uci.edu/~gohlke/pythonlibs/#panda`.

The following script opens a shapefile and dumps it into GeoJSON. Then, it creates a map with `matplotlib`:

```
>>> import geopandas
>>> import matplotlib.pyplot as plt
>>> gdf = geopandas.GeoDataFrame
>>> census = gdf.from_file("GIS_CensusTract_poly.shp")
>>> census.plot()
>>> plt.show()
```

The following image is the resulting map plot of the previous commands:

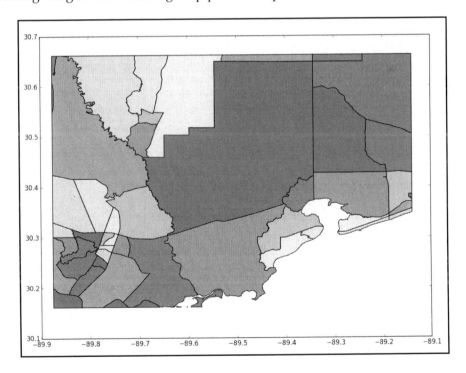

PyMySQL

The popular MySQL (available at `http://dev.mysql.com/downloads`) database is gradually evolving spatial functions. It has support for OGC geometries and a few spatial functions. It also has a pure Python API available in the PyMySQL library. The limited spatial functions use planar geometry and bounding rectangles as opposed to spherical geometry and shapes. The latest development release of MySQL contains some additional functions that improve this capability.

In the following example, we'll create a database in MySQL called `spatial_db`. Then, we'll add a table called `PLACES` with a geometry column. Next, we'll add two cities as point locations. Finally, we'll calculate the distance using MySQL's `ST_Distance` function and then convert the result from degrees into miles.

First, we will import our `mysql` library and set up the database connection:

```
# Import the python mysql library
import pymysql
# Establish a database connection on our local
# machine as the root database user.
conn = pymysql.connect(host='localhost', port=3306,
user='root', passwd='', db='mysql')
```

Next, we get the database cursor:

```
# Get the database cursor needed to change
# the database
cur = conn.cursor()
```

Now, we check if the database already exists, and drop it if it does:

```
# If the database already exists, delete
# it and recreate it either way.
cur.execute("DROP DATABASE IF EXISTS spatial_db")
cur.execute("CREATE DATABASE spatial_db")
# Close the cursor and the connection
cur.close()
conn.close()
```

Now, we set up a new connection and get a cursor:

```
# Set up a new connection and cursor
conn = pymysql.connect(host='localhost', port=3306,
user='root', passwd='', db='spatial_db')
cur = conn.cursor()
```

Next, we can create our new table and add our fields:

```
# Create our geospatial table
cur.execute("CREATE TABLE PLACES (id int NOT NULL
# Add name and location fields. The location
# field is spatially enabled to hold GIS data
AUTO_INCREMENT PRIMARY KEY, Name varchar(50) NOT NULL, location
Geometry NOT NULL)")
```

With the fields added, we are ready to insert records for the location of some cities:

```
# Insert a name and location for the city of
# New Orleans
cur.execute("INSERT INTO PLACES (name, location) VALUES ('NEW
ORLEANS', GeomFromText('POINT(30.03 90.03)'))")
# Insert a name and location for the city of
# Memphis.
cur.execute("INSERT INTO PLACES (name, location) VALUES
('MEMPHIS', GeomFromText('POINT(35.05 90.00)'))")
```

Then, we can commit changes to the database:

```
# Commit the changes to the database
conn.commit()
```

Now, we can query the database! First, we'll get a list of all of the point locations:

```
# Now let's read the data. Select all of
# the point locations from the database.
cur.execute("SELECT AsText(location) FROM PLACES")
```

Now, we'll extract the two points from the query results:

```
# We know there's only two points, so we'll
# just parse them.
p1, p2 = [p[0] for p in cur.fetchall()]
```

Before we can measure the distance, we need to convert the point listings into geospatial geometries:

```
# Now we'll convert the data
# to geometries to measure the distance
# between the two cities
cur.execute("SET @p1 = ST_GeomFromText('{}')".format(p1))
cur.execute("SET @p2 = ST_GeomFromText('{}')".format(p2))
```

Finally, we can use the `Distance` stored procedure to measure the distance between the two geometries:

```
# Now we do the measurement function which
# is also a database query.
cur.execute("SELECT ST_Distance(@p1, @p2)")
d = float(cur.fetchone()[0])
# Print the distance as a formatted
# string object.
print("{:.2f} miles from New Orleans to Memphis".format(d *
70))
cur.close()
conn.close()
```

The output is as follows:

351.41 miles from New Orleans to Memphis

There are other spatial database options available, including PostGIS and SpatiaLite; however, Python 3 support for these spatial engines is developmental at best. You can access PostGIS and MySQL through the OGR library; however, MySQL support is limited.

PyFPDF

The pure Python PyFPDF library is a lightweight way to create PDFs, including maps. Because the PDF format is a widely used standard, PDFs are commonly used to distribute maps. You can install it via PyPI as `fpdf`. The official name of the software is PyFPDF because it is a part of the PHP language module called `fpdf`. This module uses a concept called a cell to lay items out at specific locations on a page. As a quick example, we'll import the `hancock.png` image we created from the PIL example into a PDF called `map.pdf` to create a simple PDF map. The map will have the header text at the top that says Hancock County Boundary, followed by the map image:

```
>>> import fpdf
>>> # PDF constructor:
>>> # Portrait, millimeter units, A4 page size
>>> pdf=fpdf.FPDF("P", "mm", "A4")
>>> # create a new page
>>> pdf.add_page()
>>> # Set font: arial, bold, size 20
>>> pdf.set_font('Arial','B',20)
>>> # Layout cell: 160 x 25mm, title, no border, centered
```

```
>>> pdf.cell(160,25,'Hancock County Boundary', \
>>> border=0, align="C")
>>> # Write the image specifying the size
>>> pdf.image("hancock.png",25,50,110,160)
>>> # Save the file: filename, F = to file System
>>> pdf.output('map.pdf','F')
```

If you open the PDF file named map.pdf in Adobe Acrobat Reader or another PDF reader such as Sumatra PDF, you'll see that the image is now centered on an A4 page. Geospatial products are often included as part of larger reports, and the PyFPDF module simplifies automatically generating reports as PDFs.

Geospatial PDF

The **Portable Document Format**, or **PDF**, is a file format for storing and presenting digitally-formatted text and images in a cross-platform and application-independent way. PDF is a widely used document format that has also been extended to store geospatial information.

The PDF specification, starting with version 1.7, includes extensions for geospatial PDFs that map portions of the document to a physical space, also known as georeferencing. You can create points, lines, or polygons as geospatial geometries, which can also have attributes.

There are two methods for encoding geospatial information within a PDF. A company named TerraGo created a specification that has been adopted by the Open Geospatial Consortium as a best practice which is not a standard. That format is known as **GeoPDF**. The extensions that were proposed by Adobe Systems, which created the PDF specification known as ISO 32000, are currently being incorporated into the 2.0 version of the specification.

The geospatial PDF products by TerraGo conform to the OGC best practice document and the Adobe PDF extension. But TerraGo goes beyond those features to include layers and other GIS functionality. However, you must use TerraGo's plugins for Adobe Acrobat or other software to access that functionality. At a minimum, TerraGo supports the features that are needed to at least display in any PDF software.

In Python, there is a library called geopdf which has nothing to do with TerraGo but does support the OGC best practice. This library was originally developed by Tyler Garner of Prominent Edge (https://prominentedge.com/) for Python 2. It has been ported to Python 3.

Installing `geopdf` from GitHub is as simple as running the following:

```
pip install
https://github.com/GeospatialPython/geopdf-py3/archive/master.zip
```

The following example recreates the map we created in Chapter 1, *Learning about Geospatial Analysis with Python,* in the *Simple GIS* section as a geospatial PDF. The `geopdf` library relies on the Python ReportLab PDF library. The steps we will need to execute are as follows:

1. Create a PDF drawing canvas.
2. Draw a rectangle for the state of Colorado.
3. Set up a function to convert map coordinates into screen coordinates.
4. Draw and label the cities and populations.
5. Register the corners of the state as geospatial PDF coordinates that georeference the entire map.

The Python code's comments explain what's happening in each step:

```python
# Import the geopdf library
from geopdf import GeoCanvas
# Import the necessary Reportlab modules
from reportlab.pdfbase.pdfdoc import PDFString, PDFArray
# Create a canvas with a name for our pdf.
canvas = GeoCanvas('SimpleGIS.pdf')
# Draw a rectangle to represent the State boundary
canvas.rect(100, 400, 400, 250, stroke=1)
# DATA MODEL
# All layers will have a name, 1+ points, and population count
NAME = 0
POINTS = 1
POP = 2
# Create the state layer
state = ["COLORADO", [[-109, 37], [-109, 41], [-102, 41], [-102, 37]],
5187582]
# Cities layer list
# city = [name, [point], population]
cities = []
# Add Denver
cities.append(["DENVER", [-104.98, 39.74], 634265])
# Add Boulder
cities.append(["BOULDER", [-105.27, 40.02], 98889])
# Add Durango
cities.append(["DURANGO", [-107.88, 37.28], 17069])
# MAP GRAPHICS RENDERING
map_width = 400
```

```
map_height = 250
# State Bounding Box
# Use Python min/max function to get state bounding box
minx = 180
maxx = -180
miny = 90
maxy = -90
for x, y in state[POINTS]:
    if x < minx:
        minx = x
    elif x > maxx:
        maxx = x
    if y < miny:
        miny = y
    elif y > maxy:
        maxy = y
# Get earth distance on each axis
dist_x = maxx - minx
dist_y = maxy - miny
# Scaling ratio each axis
# to map points from world to screen
x_ratio = map_width / dist_x
y_ratio = map_height / dist_y
def convert(point):
    """Convert lat/lon to screen coordinates"""
    lon = point[0]
    lat = point[1]
    x = map_width - ((maxx - lon) * x_ratio)
    y = map_height - ((maxy - lat) * y_ratio)
    # Python turtle graphics start in the middle of
    # the screen so we must offset the points so they
    # are centered
    x = x + 100
    y = y + 400
    return [x, y]

# Set up our map labels
canvas.setFont("Helvetica", 20)
canvas.drawString(250, 500, "COLORADO")

# Use smaller text for cities
canvas.setFont("Helvetica", 8)

# Draw points and label the cities
for city in cities:
pixel = convert(city[POINTS])
print(pixel)
```

```
# Place a point for the city
canvas.circle(pixel[0], pixel[1], 5, stroke=1, fill=1)

# Label the city
canvas.drawString(pixel[0] + 10, pixel[1], city[NAME] + ", Population: " +
str(city[POP]))

# A series of registration point pairs (pixel x,
# pixel y, x, y) to spatially enable the PDF. We only
# need to do the state boundary.
# The cities will be contained with in it.
registration = PDFArray([
PDFArray(map(PDFString, ['100', '400', '{}'.format(minx),
'{}'.format(maxy)])),
PDFArray(map(PDFString, ['500', '400', '{}'.format(maxx),
'{}'.format(maxy)])),
PDFArray(map(PDFString, ['100', '150', '{}'.format(minx),
'{}'.format(miny)])),
PDFArray(map(PDFString, ['500', '150', '{}'.format(maxx),
'{}'.format(miny)]))
])
# Add the map registration
canvas.addGeo(Registration=registration)
# Save our geopdf
canvas.save()
```

Rasterio

The GDAL library we introduced earlier in this chapter is extremely powerful, but it wasn't designed for Python. The `rasterio` library solves that problem by wrapping GDAL in a very simple, clean Pythonic API for raster data operations.

This example uses the satellite image from the GDAL example in this chapter. We'll open the image and get some metadata, like the following

```
>>> import rasterio
>>> ds = rasterio.open("SatImage.tif")
>>> ds.name
'SatImage.tif'
>>> ds.count
3
>>> ds.width
2592
>>> ds.height
2693
```

OSMnx

The `osmnx` library combines **Open Street Map (OSM)** and the powerful NetworkX library to manage street networks used for routing. This library has dozens of dependencies which it rolls up to do all of the complex steps of downloading, analyzing, and visualizing street networks.

You can try to install `osmnx` using `pip`:

```
pip install osmnx
```

However, you may run into some installation issues due to the dependencies. In that case, it's easier to use the Conda system, which we'll introduce later in this chapter.

The following example uses `osmnx` to download street data from OSM for a city, creates a street network from it, and calculates some basic statistics:

```
>>> import osmnx as ox
>>> G = ox.graph_from_place('Bay Saint Louis, MS , USA',
network_type='drive')
>>> stats = ox.basic_stats(G)
>>> stats["street_length_avg"]
172.1468804611654
```

Jupyter

The Jupyter project is something you should be aware of when working with geospatial or other scientific data. The Jupyter Notebook app creates and displays notebook documents in a web browser that are human-readable and machine-executable code and data. It's great for sharing tutorials for software and has become very common in the geospatial Python world.

You can find a good introduction for Jupyter Notebooks and Python here: `https:// jupyter-notebook-beginner-guide.readthedocs.io/en/latest/what_is_jupyter.html`.

Conda

Conda is an open source package management system that makes installing and updating complex libraries easier. It works with several languages, including Python. Conda is very useful for setting up libraries and testing them so that we can try out new things in a development environment. It's usually better to custom configure production environments, but Conda is a great way to prototype new ideas.

You can get started with Conda at `https://conda.io/en/latest/`.

Summary

In this chapter, we surveyed the Python-specific tools for geospatial analysis. Many of these tools included bindings to the libraries we discussed in `Chapter 3`, *The Geospatial Technology Landscape,* for best-of-breed solutions for specific operations such as GDAL's raster access functions. We also included pure Python libraries as much as possible and will continue to include pure Python algorithms as we work through the upcoming chapters.

In the next chapter, we'll begin applying all of these tools for GIS analysis.

Further reading

The following links will allow you to explore the topics in this chapter further. The first link is about the XPath query language, which we used to filter XML elements using Elementree. The second link is the documentation for the Python string library, which will be critical throughout this book for manipulating data. Third, we have the `lxml` library, one of the more powerful and fast XML libraries. Finally, we have Conda, which provides a comprehensive, easy-to-use framework for scientific operations in Python, including geospatial technology:

- For more information on XPath, check out the following link: `http://www.w3schools.com/xsl/xpath_intro.asp`
- For more details on the Python `string` module, check out the following link: `https://docs.python.org/3.4/library/string.html`
- The documentation on LXML can be found at the following link: `http://lxml.de/`
- You can learn more about Conda at the following link: `https://conda.io/en/latest/`

5
Python and Geographic Information Systems

This chapter will focus on applying Python to functions that are typically performed by a **geographic information system** (**GIS**) such as QGIS or ArcGIS. These functions are the heart and soul of geospatial analysis. We will continue to use as few external dependencies as possible outside of Python itself so that you have tools that are as reusable as possible in different environments. In this book, we separate GIS analysis and remote sensing from a programming perspective, which means that, in this chapter, we'll mostly focus on vector data.

As with the other chapters in this book, the items presented here are core functions that serve as building blocks that you can recombine to solve challenges that you will encounter beyond this book. The topics in this chapter include the following:

- Measuring distance
- Converting coordinates
- Reprojecting vector data
- Measuring area
- Editing shapefiles
- Selecting data from within larger datasets
- Creating thematic maps
- Using spreadsheets
- Conversion of non-GIS data types
- Geocoding
- Multiprocessing

This chapter contains many code samples. In addition to the text, code comments are included as guides within the samples. This chapter covers more ground than any other chapter in this book. It covers everything from measuring the earth to editing data and creating maps, to using scaled up multiprocessing for faster analysis. By the end of this chapter, you'll be a geospatial analyst ready to learn about the more advanced techniques in the rest of this book.

Technical requirements

For this chapter, you will require the following:

- Python 3.7
- The Python UTM library
- The Python OGR library
- The Python Shapefile library
- The Python Fiona library
- The Python PNGCanvas library
- The Python Pillow library (Python Imaging Library)
- The Python Folium library
- The Python Pymea library
- The Python Geocoder library
- The Python GeoPy library

Measuring distance

The essence of geospatial analysis is discovering the relationships of objects on Earth. Items that are closer together tend to have a stronger relationship than those that are farther apart. This concept is known as **Tobler's First Law of Geography**. Therefore, measuring distance is a critical function of geospatial analysis.

As we have learned, every map is a model of the Earth and they are all wrong to some degree. For this reason, measuring the accurate distance between two points on the Earth while sitting in front of a computer is impossible. Even professional land surveyors (who go out in the field with both traditional sighting equipment and very precise GPS equipment) fail to account for every anomaly in the Earth's surface between point A and point B. So, to measure distance, we must look at the following questions:

- What are we measuring?
- How much are we measuring?
- How much accuracy do we need?

Now, to calculate distance, there are three models of the Earth that we can use:

- Flat plane
- Spherical
- Ellipsoid

In the flat plane model, standard Euclidean geometry is used. The Earth is considered a flat plane with no curvature, as shown in the following diagram:

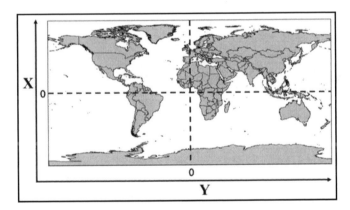

This model makes math quite simple because you work with straight lines. The most common format for geospatial coordinates is decimal degrees. However, decimal degree coordinates are reference measurements on a sphere taken as angles – between the longitude and the prime meridian—and the latitude and equator. Furthermore, the lines of longitude converge toward zero at the poles. The circumference of each line of latitude becomes smaller toward the poles as well. These facts mean decimal degrees are not a valid coordinate system for Euclidean geometry, which uses infinite planes.

Map projections attempt to simplify the issues of dealing with a 3D ellipsoid in a 2D plane, either on paper or on a computer screen. As we discussed in Chapter 1, *Learning about Geospatial Analysis with Python*, map projections flatten a round model of the Earth to a plane and introduce distortion in exchange for the convenience of a map. Once this projection is in place and decimal degrees are traded for a Cartesian coordinate system with x and y coordinates, we can use the simplest forms of Euclidean geometry—namely, the Pythagorean theorem.

At a large enough scale, a sphere or ellipsoid like the Earth appears more like a plane than a sphere. In fact, for centuries, everyone thought the Earth was flat! If the difference in degrees of longitude is small enough, you can often get away with using Euclidean geometry and then converting the measurements into meters, kilometers, or miles. This method is generally not recommended but the decision is ultimately up to you and your requirements for accuracy as an analyst.

The spherical model approach tries to better approximate reality by avoiding the problems resulting from smashing the Earth onto a flat surface. As the name suggests, this model uses a perfect sphere for representing the Earth (similar to a physical globe), which allows us to work with degrees directly. This model ignores the fact that the Earth is really more of an egg-shaped ellipsoid with varying degrees of thickness in its crust. But by working with distance on the surface of a sphere, we can begin to measure longer distances with more accuracy. The following screenshot illustrates this concept:

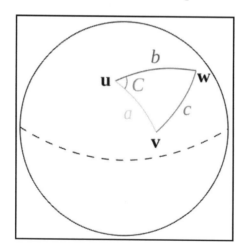

Using the ellipsoid model of the Earth, analysts strive for the best model of the Earth's surface. There are several ellipsoid models, which are called datums. A **datum** is a set of values that define an estimated shape for the Earth, also known as a **geodetic system**. Like any other georeferencing system, a datum can be optimized for a localized area. The most commonly used datum is called **WGS84**, which is designed for global use. You should be aware that WGS84 is occasionally updated as assessment techniques and technology improves. The most recent revision occurred in 2004.

In North America, the NAD83 datum is used to optimize referencing over the continent. In the Eastern Hemisphere, the **European Terrestrial Reference System 1989 (ETRS89)** is used more frequently. ETRS89 is fixed to the stable part of the **Eurasian Plate**. Maps of Europe based on ETRS89 are immune to continental drift, which changes up to 2.5 cm per year as the Earth's crust shifts.

An ellipsoid does not have a constant radius from the center. This fact means the formulas used in the spherical model of the Earth begin to have issues in the ellipsoid model. Though not a perfect approximation, it is much closer to reality than the spherical model.

The following screenshot shows a generic ellipsoid model denoted by a black line contrasted against a representation of the Earth's uneven crust, which is using a red line to represent the geoid. Although we will not use it for these examples, another model is the geoid model. The geoid is the most precise and accurate model of the Earth, which is based on the Earth's surface with no influencing factors except gravity and rotation. The following diagram is a representation of a geoid, ellipsoid, and spherical model to illustrate their differences:

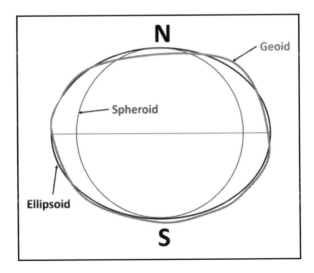

Understanding these models of the Earth is critical to everything else in this book because we're modeling the Earth, after all.

Now that we've discussed these different models of the Earth and the issues with measuring them, let's look at some solutions using Python.

Using the Pythagorean theorem

We'll start by measuring with the simplest method, that is, the Pythagorean theorem, also known as Euclidean distance. If you remember your geometry lessons from school, the Pythagorean theorem asserts the following:

$$a^2 + b^2 = c^2$$

In this assertion, the variables *a*, *b*, and *c* are all sides of a triangle. You can solve any one side if you know the other two.

In this example, we'll start with two projected points in the **Mississippi Transverse Mercator (MSTM)** projection. The units of this projection are in meters. The *x*-axis locations are measured from the central meridian defined by the westernmost location in the state. The *y*-axis is defined from the NAD83 horizontal datum. The first point, defined as ($x1$,$y1$), represents Jackson, the state capital of Mississippi. The second point, defined as ($x2$,$y2$) represents the city of Biloxi, which is a coastal town, as shown in the following illustration:

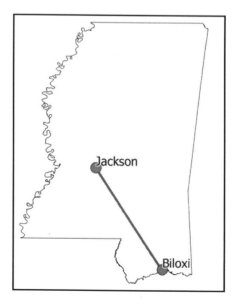

In the following example, the double-asterisk (**) in Python is the syntax for exponents, which we'll use to square the distances.

We'll import the Python math module for its square root function called `sqrt()`. Then, we'll calculate the *x*-axis and *y*-axis distances. Finally, we'll use these variables to execute the Euclidean distance formula to get the distance across the bounding box in meters from an *x, y* origin, which will be used in the MSTM projection:

```python
import math
# First point
x1 = 456456.23
y1 = 1279721.064
# Second point
x2 = 576628.34
y2 = 1071740.33
# X distance
x_dist = x1 - x2
# Y distance
y_dist = y1 - y2
# Pythagorean theorem
dist_sq = x_dist**2 + y_dist**2
distance = math.sqrt(dist_sq)
print(distance)
# 240202.66
```

So, the distance is approximately 240,202 meters, which is around 240.2 kilometers or 150 miles. This calculation is reasonably accurate because this projection is optimized for measuring distance and area in Mississippi using Cartesian coordinates.

We can also measure distance using decimal degrees, but we must perform a few additional steps. To measure using degrees, we must convert the angles into radians, which account for the curved surface distance between the coordinates. We'll also multiply our output in radians times the radius of the Earth in meters to convert back from radians.

You can read more about radians at `http://en.wikipedia.org/wiki/Radian`.

We'll perform this conversion using the Python `math.radians()` method in the following code when we calculate the *x* and *y* distances:

```
import math
x1 = -90.21
y1 = 32.31
x2 = -88.95
y2 = 30.43
x_dist = math.radians(x1 - x2)
y_dist = math.radians(y1 - y2)
dist_sq = x_dist**2 + y_dist**2
dist_rad = math.sqrt(dist_sq)
dist_rad * 6371251.46
# 251664.46
```

Okay, so this time, we came up with around 251 kilometers, which is 11 kilometers more than our first measurement. So, as you can see, your choice of measurement algorithm and Earth model can have significant consequences. Using the same equation, we come up with radically different answers, depending on our choice of coordinate system and Earth model.

 You can read more about Euclidean distance at `http://mathworld.wolfram.com/Distance.html`.

Let's check out the haversine formula next.

Using the haversine formula

Part of the problem with using the Pythagorean theorem to measure distance on the Earth, which is a sphere, is the concept of great circle distance. A great circle is the shortest distance between two points on a sphere. Another important feature that defines a great circle is that the circle, if followed all of the way around the sphere, will bisect the sphere into two equal halves, as shown in the following Wikipedia illustration:

So, what is the right way to measure a line on a curved sphere? The most popular method is to use the **haversine formula**, which uses trigonometry to calculate the Great Circle distance using coordinates defined in decimal degrees as input. The haversine formula is $haversine(\theta) = sin^2(\theta/2)$, where θ is the central angle between two points on a sphere. Once again, we'll convert the axis distances from degrees into radians before we apply the formula, just like in the previous example. But this time, we'll also convert the latitude (*y*-axis) coordinates into radians separately:

```python
import math
x1 = -90.212452861859035
y1 = 32.316272202663704
x2 = -88.952170968942525
y2 = 30.438559624660321
x_dist = math.radians(x1 - x2)
y_dist = math.radians(y1 - y2)
y1_rad = math.radians(y1)
y2_rad = math.radians(y2)
a = math.sin(y_dist/2)**2 + math.sin(x_dist/2)**2 \
  * math.cos(y1_rad) * math.cos(y2_rad)
c = 2 * math.asin(math.sqrt(a))
distance = c * 6371   # kilometers
print(distance)
# 240.63
```

Wow! We get 240.6 kilometers using the haversine formula, compared to 240.2 kilometers using the optimized and more accurate projection. This difference is less than half a kilometer, which is not bad for a distance calculation of two cities 150 miles apart. The haversine formula is the most commonly used distance measuring formula because it is relatively lightweight from a coding perspective and reasonably accurate in most cases. It is considered to be accurate to within about a meter.

To summarize what we've learned so far, most of the point coordinates you encounter as an analyst are in unprojected decimal degrees. So, your options for measurement are as follows:

- Reproject to a distance-accurate Cartesian projection and measure.
- Just use the haversine formula and see how far it takes you for your analysis.
- Use the even more precise Vincenty formula.

That's right! There's another formula that seeks to provide an even better measurement than haversine.

Using the Vincenty formula

So, we've examined distance measurement using the Pythagorean theorem (flat Earth model) and the haversine formula (spherical Earth model). The Vincenty formula accounts for the ellipsoid model of the Earth. And if you are using a localized ellipsoid, it can be accurate to much less than a meter.

In the following implementation of this formula, you can change the semi-major axis value and flattening ratio to fit the definition of any ellipsoid. Let's see what the distance is when we measure using the Vincenty formula on the NAD83 ellipsoid in the following example:

1. First, we will import the `math` module, which allows us to work in radians, and the other `math` functions we'll need:

   ```
   import math
   ```

2. Now, we need to set up our variables, including the variable that holds our distance value, the two points we're measuring, the constants describing the Earth, and the derivative formulas we need:

   ```
   distance = None
   x1 = -90.212452861859035
   y1 = 32.316272202663704
   x2 = -88.952170968942525
   y2 = 30.438559624660321
   ```

```
# Ellipsoid Parameters
# Example is NAD83
a = 6378137  # semi-major axis
f = 1/298.257222101  # inverse flattening
b = abs((f*a)-a)  # semi-minor axis
L = math.radians(x2-x1)
U1 = math.atan((1-f) * math.tan(math.radians(y1)))
U2 = math.atan((1-f) * math.tan(math.radians(y2)))
sinU1 = math.sin(U1)
cosU1 = math.cos(U1)
sinU2 = math.sin(U2)
cosU2 = math.cos(U2)
lam = L
```

3. Now begins the Vincenty formula. There's just no easy way to do this and the math is a little complicated, but it works:

```
for i in range(100):
    sinLam = math.sin(lam)
    cosLam = math.cos(lam)
    sinSigma = math.sqrt((cosU2*sinLam)**2 +
                    (cosU1*sinU2-sinU1*cosU2*cosLam)**2)
    if (sinSigma == 0):
        distance = 0  # coincident points
        break
    cosSigma = sinU1*sinU2 + cosU1*cosU2*cosLam
    sigma = math.atan2(sinSigma, cosSigma)
    sinAlpha = cosU1 * cosU2 * sinLam / sinSigma
    cosSqAlpha = 1 - sinAlpha**2
    cos2SigmaM = cosSigma - 2*sinU1*sinU2/cosSqAlpha
    if math.isnan(cos2SigmaM):
        cos2SigmaM = 0  # equatorial line
    C = f/16*cosSqAlpha*(4+f*(4-3*cosSqAlpha))
    LP = lam
    lam = L + (1-C) * f * sinAlpha *
        (sigma + C*sinSigma*(cos2SigmaM+C*cosSigma *
                        (-1+2*cos2SigmaM*cos2SigmaM)))
    if not abs(lam-LP)  1e-12:
        break
uSq = cosSqAlpha * (a**2 - b**2) / b**2
A = 1 + uSq/16384*(4096+uSq*(-768+uSq*(320-175*uSq)))
B = uSq/1024 * (256+uSq*(-128+uSq*(74-47*uSq)))
deltaSigma = B*sinSigma*(cos2SigmaM+B/4 *
(cosSigma*(-1+2*cos2SigmaM*cos2SigmaM) -
B/6*cos2SigmaM*(-3+4*sinSigma*sinSigma) *
(-3+4*cos2SigmaM*cos2SigmaM)))
s = b*A*(sigma-deltaSigma)
```

Finally, after all that, we have our distance:

```
distance = s
print(distance)
# 240237.66693880095
```

Using the Vincenty formula, our measurement came to 240.1 kilometers, which is only 100 meters off from our projected measurement using Euclidean distance. Impressive! While many times more mathematically complex than the haversine formula, you can see that it is also much more accurate.

 The pure Python geopy module includes an implementation of the Vincenty formula and has the ability to geocode locations by turning place names into latitude and longitude coordinates: http://geopy. readthedocs.org/en/latest/.

The points that were used in these examples are reasonably close to the equator. As you move toward the poles or work with larger distances or extremely small distances, the choices you make become increasingly more important. If you're just trying to make a radius around a city to select locations for a marketing campaign promoting a concert, then an error of a few kilometers is probably okay. However, if you're trying to estimate fuel required for an airplane to make a flight between two airports, then you want to be spot on!

If you'd like to learn more about issues with measuring distance and direction, and how to work around them with programming, visit the following site: http://www.movable-type. co.uk/scripts/latlong.html.

On this site, Chris Veness goes into great detail on this topic and provides online calculators, as well as examples written in JavaScript, which can easily be ported to Python. The Vincenty formula implementation that we just saw is ported from the JavaScript on this site.

You can see the full pure mathematical notation for the Vincenty formula here: https:// en.wikipedia.org/wiki/Vincenty%27s_formulae.

Now that we know how to calculate distance, we need to understand how to calculate the direction of a line to relate objects on Earth by distance and location for geospatial analysis.

Calculating line direction

In addition to distance, you will often want to know the bearing of a line between its endpoints. We can calculate this line direction from one of the points using only the Python `math` module:

1. First, we import the `math` functions we'll need:

```
from math import atan2, cos, sin, degrees
```

2. Next, we set up some variables for our two points:

```
lon1 = -90.21
lat1 = 32.31
lon2 = -88.95
lat2 = 30.43
```

3. Next, we'll calculate the angle between the two points:

```
angle = atan2(cos(lat1)*sin(lat2)-sin(lat1) * \
    cos(lat2)*cos(lon2-lon1), sin(lon2-lon1)*cos(lat2))
```

4. Finally, we'll calculate the bearing of the line in degrees:

```
bearing = (degrees(angle) + 360) % 360
print(bearing)
309.3672990606595
```

Sometimes, you end up with a negative bearing value. To avoid this issue, we add `360` to the result to avoid a negative number and use the Python modulo operator to keep the value from climbing to over `360`.

The `math` in the angle calculation is reverse engineering a right triangle and then figuring out the acute angle of the triangle. The following URL provides an explanation of the elements of this formula, along with an interactive example at the end: `https://www.mathsisfun.com/sine-cosine-tangent.html`.

We now know how to calculate the location of features on the Earth. Next, we'll learn how to integrate data from different sources, starting with coordinate conversion.

Understanding coordinate conversion

Coordinate conversion allows you to convert point coordinates between different coordinate systems. When you start working with multiple datasets, you'll inevitably end up with data in different coordinate systems and projections. You can convert back and forth between two of the most common coordinate systems, UTM and geographic coordinates (latitude and longitude), using a pure Python module called utm. You can install it using `easy_install` or `pip` from PyPI: `https://pypi.python.org/pypi/utm`.

The utm module is straightforward to use. To convert from UTM into latitude and longitude, you can use the following code:

```
import utm
y = 479747.0453210057
x = 5377685.825323031
zone = 32
band = 'U'
print(utm.to_latlon(y, x, zone, band))
# (48.55199390882121, 8.725555729071763)
```

The UTM zones are numbered horizontally. However, vertically, the bands of latitude are ordered by the English alphabet with a few exceptions. For example, the letters *A, B, Y*, and *Z* are used to label the Earth's poles. The letters *I* and *O* are omitted because they look too much like *1* and *0*. Letters *N* through *X* are in the Northern Hemisphere while *C* through *M* are in the Southern Hemisphere. The following screenshot, from the website *Atlas Florae Europaeae*, illustrates the UTM zones over Europe:

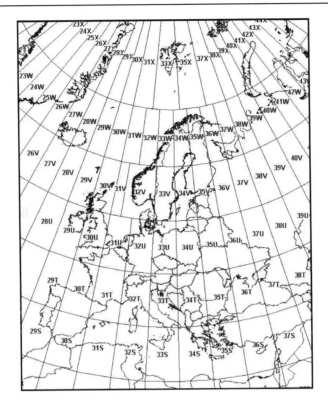

Converting from latitude and longitude is even easier. We just pass the latitude and longitude to the `from_latlon()` method, which returns a tuple with the same parameters that are accepted by the `to_latlon()` method:

```
import utm
utm.from_latlon(48.55199390882121, 8.725555729071763)
# (479747.04524576373, 5377691.373080335, 32, 'U')
```

The algorithms that were used in this Python implementation are described in detail at `http://www.uwgb.edu/dutchs/UsefulData/` `UTMFormulas.HTM`.

Converting between UTM and latitude/longitude just scratches the surface of transforming datasets from different sources so that they can be overlaid nicely on a map. To go beyond the basics, we'll need to perform map projections.

Now that we know how to calculate line direction, let's see how reprojection is done.

Understanding reprojection

In GIS, reprojection is all about changing the coordinates in a dataset from one coordinate system to another. While reprojection is less common these days due to more advanced methods of data distribution, sometimes you need to reproject a shapefile. The pure Python utm module works for reference system conversion, but for a full reprojection, we need some help from the OGR Python API. The OGR API contained in the osgeo module also provides the Open Spatial Reference module, also known as osr, which we'll use for reprojection.

As an example, we'll use a point shapefile containing New York City museum and gallery locations in the Lambert conformal projection. We'll reproject it to WGS84 geographic (or un-project, it rather). You can download this zipped shapefile at https://git.io/vLbT4.

The following minimalist script reprojects the shapefile. The geometry is transformed and then written to the new file, but the .dbf file is simply copied to the new name as we aren't changing it. The standard Python shutil module, short for shell utilities, is used to copy .dbf. The source and target shapefile names are variables at the beginning of the script. The target projection is also near the top, which is set using an EPSG code. The script assumes there is a .prj projection file, which defines the source projection. If not, you could manually define it using the same syntax as the target projection. We'll walk through projecting a dataset step by step. Each section is marked with comments:

1. First, we import our libraries:

```
from osgeo import ogr
from osgeo import osr
import os
import shutil
```

2. Next, we define our shapefile names as variables:

```
srcName = 'NYC_MUSEUMS_LAMBERT.shp'
tgtName = 'NYC_MUSEUMS_GEO.shp'
```

3. Now, we create our target spatial reference using the osr module as EPSG code 4326, which is WGS84 Geographic:

```
tgt_spatRef = osr.SpatialReference()
tgt_spatRef.ImportFromEPSG(4326)
```

4. Then, we set up our shapefile `Reader` object using `ogr` and get the spatial reference:

```
driver = ogr.GetDriverByName('ESRI Shapefile')
src = driver.Open(srcName, 0)
srcLyr = src.GetLayer()
src_spatRef = srcLyr.GetSpatialRef()
```

5. Next, we check whether our target shapefile already exists from a previous test run and delete it if it does:

```
if os.path.exists(tgtName):
    driver.DeleteDataSource(tgtName)
```

6. Now, we can begin building our target layer for the shapefile:

```
tgt = driver.CreateDataSource(tgtName)
lyrName = os.path.splitext(tgtName)[0]
# Use well-known binary format (WKB) to specify geometry
tgtLyr = tgt.CreateLayer(lyrName, geom_type=ogr.wkbPoint)
featDef = srcLyr.GetLayerDefn()
trans = osr.CoordinateTransformation(src_spatRef, tgt_spatRef)
```

7. Next, we can loop through the features in our source shapefile, reproject them using the `Transform()` method, and add them to the new shapefile:

```
srcFeat = srcLyr.GetNextFeature()
while srcFeat:
    geom = srcFeat.GetGeometryRef()
    geom.Transform(trans)
    feature = ogr.Feature(featDef)
    feature.SetGeometry(geom)
    tgtLyr.CreateFeature(feature)
    feature.Destroy()
    srcFeat.Destroy()
    srcFeat = srcLyr.GetNextFeature()
src.Destroy()
tgt.Destroy()
```

8. Then, we need to create a shapefile `.prj` file containing projection information as a shapefile has no inherent way to store it:

```
# Convert geometry to Esri flavor of Well-Known Text (WKT) format
# for export to the projection (prj) file.
tgt_spatRef.MorphToESRI()
prj = open(lyrName + '.prj', 'w')
prj.write(tgt_spatRef.ExportToWkt())
prj.close()
```

9. Finally, we can just make a copy of the `.dbf` source with the new filename as the attributes are part of the reprojection process:

```
srcDbf = os.path.splitext(srcName)[0] + '.dbf'
tgtDbf = lyrName + '.dbf'
shutil.copyfile(srcDbf, tgtDbf)
```

The following screenshot shows the reprojected points in QGIS with satellite imagery in the background:

If you are working with a set of points, you can reproject them programmatically instead of reprojecting a shapefile using PyProj: `https://jswhit.github.io/pyproj/`.

In addition to converting coordinates into different projections, you often need to convert them among different formats, which we'll look at next.

Understanding coordinate format conversion

Map coordinates were traditionally represented as degrees, minutes, and seconds (DMS) for maritime navigation. However, in GIS (which is computer-based), latitude and longitude are represented as decimal numbers known as **decimal degrees**. The degrees, minutes, and seconds format is still used. Sometimes, you have to convert between that format and decimal degrees to perform calculations and output reports.

In this example, we'll create two functions that can convert either format into the other:

1. First, we import the `math` module to do conversions and the `re` regular expression module to parse the coordinate string:

```
import math
import re
```

2. We have our function to convert decimal degrees into a `degrees`, `minutes`, and `seconds` string:

```
def dd2dms(lat, lon):
    """Convert decimal degrees to degrees, minutes, seconds"""
    latf, latn = math.modf(lat)
    lonf, lonn = math.modf(lon)
    latd = int(latn)
    latm = int(latf * 60)
    lats = (lat - latd - latm / 60) * 3600.00
    lond = int(lonn)
    lonm = int(lonf * 60)
    lons = (lon - lond - lonm / 60) * 3600.00
    compass = {
        'lat': ('N','S'),
        'lon': ('E','W')
    }
    lat_compass = compass['lat'][0 if latd >= 0 else 1]
    lon_compass = compass['lon'][0 if lond >= 0 else 1]
    return '{}º {}\' {:.2f}" {}, {}º {}\' {:.2f}" {}'.format(abs(latd),
    abs(latm), abs(lats), lat_compass, abs(lond),
    abs(lonm), abs(lons), lon_compass)
```

3. Next, we have our function to go the other way and convert degrees:

```
def dms2dd(lat, lon):
    lat_deg, lat_min, \
    lat_sec, lat_dir = re.split('[^\d\.A-Z]+', lat)
    lon_deg, lon_min, \
    lon_sec, lon_dir = re.split('[^\d\.A-Z]+', lon)
    lat_dd = float(lat_deg) +\
    float(lat_min)/60 + float(lat_sec)/(60*60);
    lon_dd = float(lon_deg) +\
    float(lon_min)/60 + float(lon_sec)/(60*60);
    if lat_dir == 'S':
        lat_dd *= -1
    if lon_dir == 'W':
        lon_dd *= -1
    return (lat_dd, lon_dd);
```

4. Now, if we want to convert decimal degrees into DMS, it's as simple as using the following code:

```
print(dd2dms(35.14953, -90.04898))
# 35º 8' 58.31" N, 90º 2' 56.33" W
```

5. To go the other direction, you just type the following function:

```
dms2dd("""29º 56' 0.00" N""", """90º 4' 12.36" W""")
(29.933333333333334, -90.0701)
```

Note that, because the DMS coordinates contain both single and double quotes to represent minutes and seconds, we have to use the Python string convention of using triple quotes on each latitude and longitude coordinate to contain both types of quotes so that they are parsed correctly.

Coordinates are the fundamental units of a GIS dataset. They are used to build points, lines, and polygons.

Calculating the area of a polygon

We have one more calculation before we move on to editing GIS data. The most basic unit of GIS is a point. Two points can form a line. Multiple lines that share endpoints can form a polyline, and polylines can form polygons. Polygons are used to represent everything from a house to an entire country in geospatial operations.

Calculating the area of a polygon is one of the most useful operations in GIS if we wish to understand the relative size of features. But in GIS, area calculations go beyond basic geometry. The polygon lies on the Earth, which is a curved surface. The polygon must be projected to account for that curvature.

Fortunately, there is a pure Python module simply called `area` that handles these complications for us. And because it's pure Python, you can look at the source code to see how it works. The `area` module's `area()` function accepts a GeoJSON string with a list of points that form a polygon and then returns the area. The following steps will show you how to calculate the area of a polygon:

1. You can install the `area` module using `pip`:

   ```
   pip install area
   ```

2. First, we'll import the `area` function from the `area` module:

   ```
   from area import area
   ```

3. Next, we'll create a variable called `polygon` that's contained in a GeoJSON geometry for our polygon:

   ```
   # Our points making up a polygon
   polygon =
   {"type":"Polygon","coordinates":[[[-89.324,30.312],[-89.326,30.31],
   [-89.322,30.31],[-89.321,30.311],[-89.321,30.312],[-89.324,30.312]]
   ]}
   ```

4. Now, we're able to pass the polygon points string to the area function to calculate the area:

   ```
   a = area(polygon)
   ```

5. The area that's returned is `80235.13927976067` square meters. We can then use Python's built-in `round()` function to round the long floating-point value to two decimal places to get **80235.14**:

   ```
   round(a, 2)
   ```

You now have the tools to do the math regarding the distance and size for geospatial data.

In the next section, we'll look at editing datasets in one of the most popular GIS data formats—shapefiles.

Editing shapefiles

Shapefiles are one of the most common data formats in GIS, both for exchanging data as well as performing GIS analysis. In this section, we'll learn how to work with these files extensively. In Chapter 2, *Learning Geospatial Data*, we discussed shapefiles as a format that can have many different file types associated with it. For editing shapefiles, and most other operations, we are only concerned with two file types:

- The .shp file
- The .dbf file

The .shp file contains the geometry while the .dbf file contains the attributes of the corresponding geometry. For each geometry record in a shapefile, there is one .dbf record. The records aren't numbered or identified in any way. This means that, when adding and deleting information from a shapefile, you must be careful to remove or add a record to each file type to match.

As we discussed in Chapter 4, *Geospatial Python Toolbox*, there are two libraries we can use to edit shapefiles in Python:

- One is the Python bindings to the OGR library.
- The other is the PyShp library, which is written in pure Python.

We'll use PyShp in order to stick with the *pure Python when possible* theme of this book. To install PyShp, use easy_install or pip.

To begin editing shapefiles, we'll start with a point shapefile containing cities for the state of Mississippi, which you can download as a ZIP file. Download the following file to your working directory and unzip it: http://git.io/vLbU4.

The points we are working with can be seen in the following illustration:

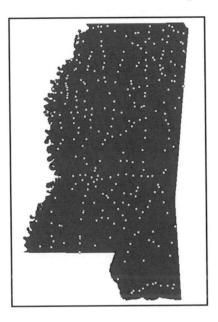

Accessing the shapefile

To do anything with a shapefile, we'll need to access it as a data source. To access the shapefile, we'll use PyShp to open it. In PyShp, we'll add the following code:

```
import shapefile
r = shapefile.Reader('MSCities_Geo_Pts')
r
<shapefile.Reader instance at 0x00BCB760>
```

We created a shapefile `Reader` object instance and set it to the `r` variable. Notice that, when we passed the filename to the `Reader` class, we didn't use any file extensions. Remember that we are dealing with at least two different files ending in `.shp` and `.dbf`. So, the base filename without the extension that is common to these two files is all we really need.

You can, however, use a file extension. PyShp will just ignore it and use the base filename. So, why would you add an extension? Most operating systems allow an arbitrary number of periods in a filename. For example, you might have a shapefile with the following base name: `myShapefile.version.1.2`.

In this case, PyShp will try to interpret the characters after the last period as a file extension, which would be .2. This issue will prevent you from opening the shapefile. So, if your shapefile has periods in the base name, you would need to add a file extension such as .shp or .dbf to the filename.

Once you have opened a shapefile and created a Reader object, you can get some information about the geographic data. In the following sample, we'll get the bounding box, shape type, and the number of records in the shapefile from our Reader object:

```
r.bbox
[-91.38804855553174, 30.29314882296931, -88.18631833931401,
34.96091138678437]
r.shapeType
# 1
r.numRecords
# 298
```

The bounding box, which is stored in the r.bbox property, is returned as a list containing the minimum x value, minimum y value, maximum x value, and maximum y value. The shape type, which is available as the shapeType property, is a numeric code defined by the official shapefile specification. In this case, 1 represents a point shapefile, 3 represents lines, and 5 represents polygons. And finally, the numRecords property tells us there are 298 records in this shapefile. Because it is a simple point shapefile, we know there are 298 points, each with their own .dbf record.

The following table shows the different geometry types for shapefiles, along with their corresponding numeric code:

Geometry	Numeric Code
NULL	0
POINT	1
POLYLINE	3
POLYGON	5
MULTIPOINT	8
POINTZ	11
POLYLINEZ	13
POLYGONZ	15
MULTIPOINTZ	18
POINTM	21
POLYLINEM	23
POLYGONM	25

MULTIPOINTM	28
MULTIPATCH	31

Now that we know how to access it, let's see how we can read these files.

Reading shapefile attributes

The .dbf file is a simple database format that is structured in a similar way to a spreadsheet with rows and columns, with each column as a label defining what information it contains. We can view that information by checking the fields property of the Reader object:

```
r.fields
# [('DeletionFlag', 'C', 1, 0), ['STATEFP10', 'C', 2, 0],
['PLACEFP10', 'C', 5, 0],
# ['PLACENS10', 'C', 8, 0], ['GEOID10', 'C', 7, 0], ['NAME10', 'C',
100, 0],
# ['NAMELSAD10', 'C', 100, 0], ['LSAD10', 'C', 2, 0], ['CLASSFP10',
'C', 2, 0],
# ['PCICBSA10', 'C', 1, 0], ['PCINECTA10', 'C', 1, 0], ['MTFCC10',
'C', 5, 0],
# ['FUNCSTAT10', 'C', 1, 0], ['ALAND10', 'N', 14, 0], ['AWATER10',
'N', 14,0],
# ['INTPTLAT10', 'C', 11, 0], ['INTPTLON10', 'C', 12, 0]]
```

The fields property returns quite a bit of information. The fields contain a list of information about each field, called **field descriptors**. For each field, the following information is presented:

- **Field name:** This is the name of the field as text, which can be no longer than 10 characters for shapefiles.
- **Field type:** This is the type of the field, which can be text, number, date, floating-point number, or Boolean represented as C, N, D, F, and L, respectively. The shapefile specification says it uses the .dbf format specified as dBASE III, but most GIS software seems to support dBASE IV. In version IV (4), the number and floating-point types are equivalent.
- **Field length:** This is the length of the data in characters or digits.
- **Decimal length:** This is the number of decimal places in a number or floating-point field.

The first field descriptor outlines a hidden field that is part of the `.dbf` file format specification. `DeletionFlag` allows the software to mark records for deletion without actually deleting them. That way, the information is still in the file but can be removed from the displayed record list or search queries.

If we just want the field name and not the other metadata, we can use Python list comprehensions to return just the first item in the descriptor and ignore the `DeletionFlag` field. This example creates a list comprehension that returns the first item in each descriptor (field name), starting with the second descriptor to ignore the deletion flag:

```
[item[0] for item in r.fields[1:]]
# ['STATEFP10', 'PLACEFP10', 'PLACENS10', 'GEOID10', 'NAME10',
'NAMELSAD10', 'LSAD10',
# 'CLASSFP10', 'PCICBSA10', 'PCINECTA10', 'MTFCC10', 'FUNCSTAT10',
'ALAND10',
# 'AWATER10', 'INTPTLAT10', 'INTPTLON10']
```

Now, we have just the field names, which are much easier to read. For clarity, the field names all contain the number `10` because this is version 2010 of this shapefile, which is created as a part of each census. These kinds of abbreviations are common in shapefile `.dbf` files due to the `10` character limit on the field names.

Next, let's examine some of the records that these fields describe. We can view an individual record using the `r.record()` method. We know from the first example that there are `298` records. So, let's examine the third record as an example. The records are accessed using list indexes. In Python, indexes start at `0`, so we have to subtract one from the desired record number to get the index. For record 3, the index would be `2`. You just pass the index to the `record()` method, as shown in the following code:

```
r.record(2)
#['28', '16620', '02406337', '2816620', 'Crosby', 'Crosby town', '43',
'C1', 'N','N', # 'G4110', 'A', 5489412, 21336, '+31.2742552',
'-091.0614840']
```

As you can see, the field names are stored separately from the actual records. If you want to select a record value, you need its index. The index of the city name in each record is 4:

```
r.record(2)[4]
# 'Crosby'
```

But counting indexes is tedious. It's much easier to reference a value by the field name. There are several ways we can associate a field name with the value of a particular record. The first is to use the `index()` method in Python lists to programmatically get the index using the field name:

```
fieldNames = [item[0] for item in r.fields[1:]]
name10 = fieldNames.index('NAME10')
name10
# 4
r.record(2)[name10]
# 'Crosby'
```

Another way we can associate field names to values is by using Python's built-in `zip()` method, which matches corresponding items in two or more lists and merges them into a list of tuples. Then, we can loop through that list, check the name, and then grab the associated value, as shown in the following code:

```
fieldNames = [item[0] for item in r.fields[1:]]
fieldNames
# ['STATEFP10', 'PLACEFP10', 'PLACENS10', 'GEOID10', 'NAME10',
'NAMELSAD10',
# 'LSAD10', 'CLASSFP10', 'PCICBSA10', 'PCINECTA10', 'MTFCC10','FUNCSTAT10',
# 'ALAND10','AWATER10', 'INTPTLAT10', 'INTPTLON10']
 rec = r.record(2)
 rec
# ['28', '16620', '02406337', '2816620', 'Crosby', 'Crosby town',
# '43', 'C1', 'N','N', 'G4110', 'A', 5489412, 21336, '+31.2742552',
'-091.0614840']
 zipRec = zip(fieldNames, rec)
 list(zipRec)
# [('STATEFP10', '28'), ('PLACEFP10', '16620'), ('PLACENS10', '02406337'),
# ('GEOID10', '2816620'), ('NAME10', 'Crosby'), ('NAMELSAD10', 'Crosby
town'),
# ('LSAD10', '43'), ('CLASSFP10', 'C1'),
('PCICBSA10','N'),('PCINECTA10','N'),
# ('MTFCC10', 'G4110'), ('FUNCSTAT10', 'A'), ('ALAND10',
5489412),('AWATER10', 21336),
# ('INTPTLAT10', '+31.2742552'), ('INTPTLON10', '-091.0614840')]
for z in zipRec:
    if z[0] == 'NAME10': print(z[1])
# Crosby
```

We can also loop through .dbf records using the r.records() method. In this example, we'll loop through the list returned by the records() method but limit the results using Python array slicing to the first three records. As we mentioned previously, shapefiles don't contain record numbers, so we'll also enumerate the records list and create a record number on the fly, so the output is a little easier to read. In this example, we'll use the enumerate() method, which will return tuples containing an index and the record, as shown in the following code:

```
for rec in enumerate(r.records()[:3]):
    print(rec[0]+1, ': ', rec[1])
# 1 :  ['28', '59560', '02404554', '2859560', 'Port Gibson', 'Port Gibson
city', '
# 25', 'C1', 'N', 'N', 'G4110', 'A', 4550230, 0, '+31.9558031',
'-090.9834329']
# 2 :  ['28', '50440', '02404351', '2850440', 'Natchez', 'Natchez city',
'25', 'C1',
#        'Y', 'N', 'G4110', 'A', 34175943, 1691489, '+31.5495016',
'-091.3887298']
# 3 :  ['28', '16620', '02406337', '2816620', 'Crosby', 'Crosby town',
'43', 'C1','N',
#        'N', 'G4110', 'A', 5489412, 21336, '+31.2742552', '-091.0614840']
```

This kind of enumeration trick is what most GIS software packages use when displaying records in a table. Many GIS analysts assume shapefiles store the record number because every GIS program displays one. But if you delete a record, for example, record number 5 in ArcGIS or QGIS, and save the file, when you open it again, you'll find what was formerly record number 6 is now record 5. Some spatial databases may assign a unique identifier to records. Often, a unique identifier is helpful. You can always create another field and column in .dbf and assign your own number, which remains constant even when records are deleted.

If you are working with very large shapefiles, PyShp has iterator methods that access data more efficiently. The default `records()` method reads all the records into the RAM at once, which is fine for the small `.dbf` files but becomes difficult to manage even with a few thousand records. Any time you'd use the `records()` method, you can also use the `r.iterRecords()` method the same way. This method holds the minimum amount of information needed to provide the record at hand rather than the whole dataset. In this quick example, we're using the `iterRecords()` method to count the number of records to verify the count in the file header:

```
counter = 0
for rec in r.iterRecords():
    counter += 1
counter
# 298
```

Now that we can read one half of the shapefile, that is, the attributes, we're ready to look at the other half, that is, the geometry.

Reading shapefile geometry

Now, let's take a look at the geometry. Previously, we looked at the header information and determined this shapefile was a point shapefile. So, we know that each record contains a single point. Let's examine the first geometry record:

```
geom = r.shape(0)
geom.points
# [[-90.98343326763826, 31.9558035947602]]
```

In each geometry record, also known as `shape`, the points are stored in a list called `points`, even if there is only one point, as in this case. Points are stored as *x*, *y* pairs, so longitude comes before latitude if that coordinate system is used.

The shapefile specification also allows for 3D shapes. Elevation values are along the *z*-axis and often called *z* values. So, a 3D point is typically described as *x*, *y*, *z*. In the shapefile format, *z* values are stored in a separate *z* attribute if they're allowed by the shape type. If the shape type doesn't allow for *z* values, then that attribute is never set when the records are read by PyShp. Shapefiles with *z* values also contain measure values or *m* values, which are rarely used and are not used in this example.

A measure is a user-assigned value that may be associated with a shape. An example would be a temperature recorded at a given location. There is another class of shape types that allow for adding *m* values to each shape but not *z* values. This class of shape types is called an **M shape type**. Just like the *z* values, if the data is there, the *m* attribute is created; otherwise, it's not. You don't typically run into shapefiles with *z* values and you rarely come across shapefiles with *m* values set. But sometimes you do, so it's good to be aware of them. And just like our fields and records .dbf example, if you don't like having the *z* and *m* values stored in separate lists, from the points list, you can use the zip() method to combine them. The zip method can take multiple lists as parameters separated by commas, as demonstrated when we looped through the records previously and joined the field names and attributes.

When you create a Reader object with PyShp, it is read-only. You can change any values in the Reader object, but they are not written to the original shapefile. In the next subsection, we'll see how we can make changes in the original shapefile.

Changing a shapefile

To create a shapefile, you need to also create a Writer object. You can change values in either a Reader or Writer object; they are just dynamic Python data types. But at some point, you must copy the values from Reader to Writer. PyShp automatically handles all of the header information, such as the bounding box and record count. You only need to worry about the geometry and attributes. You'll find that this method is much simpler than the OGR example we used previously. However, it is also limited to UTM projections.

To demonstrate this concept, we'll read in a shapefile containing points with units in degrees and convert it into the UTM reference system in a Writer object before saving it. We'll use PyShp and the UTM module we discussed previously in this chapter. The shapefile we'll use is the New York City museums shapefile, which we reprojected to a WGS84 geographic. You can also just download it as a ZIP file, which is available at https://git.io/vLd8Y.

In the following example, we'll read in the shapefile, create a writer for the converted shapefile, copy the fields over and then the records, and finally convert each point and write it as a geometry record before saving the converted shapefile:

```
import shapefile
import utm
r = shapefile.Reader('NYC_MUSEUMS_GEO')
w = shapefile.Writer(r.shapeType)
w.fields = list(r.fields)
w.records.extend(r.records())
```

```
for s in r.iterShapes():
    lon,lat = s.points[0]
    y,x,zone,band = utm.from_latlon(lat,lon)
    w.point(x,y)
w.save('NYC_MUSEUMS_UTM')
```

If you were to print out the first point of the first shape, you would see the following:

```
print(w.shapes()[0].points[0])
# [4506346.393408813, 583315.4566450359, 0, 0]
```

The point is returned as a list containing four numbers. The first two are the x and y values, while the last two are placeholders, in this case for elevation and measure values, respectively, which are used when you write those types of shapefiles. Also, we did not write a PRJ projection file, as we did in the preceding reprojection example. Here's a simple way to create a PRJ file using the EPSG code from https://spatialreference.org/. The zone variable in the preceding example tells us that we are working in UTM Zone 18, which is EPSG code 26918. The following code will create a prj file:

```
from urllib.request import urlopen
prj = urlopen('http://spatialreference.org/ref/epsg/26918/esriwkt/')
with open('NYC\_MUSEUMS\_UTM', 'w') as f:
    f.write(str(prj.read()))
```

As another example, we can add a new feature to a shapefile. In this example, we'll add a second polygon to a shapefile representing a tropical storm. You can download the zipped shapefile for this example here: https://git.io/vLdlA.

We'll read the shapefile, copy it to a `Writer` object, add the new polygon, and write it back out with the same filename using the following code:

```
import shapefile
file_name = "ep202009.026_5day_pgn.shp"
r = shapefile.Reader(file_name)
with shapefile.Writer("test", r.shapeType) as w:
    w.fields = list(r.fields)
    for rec in r.records():
        w.record(*list(rec))
    for s in r.shapes():
        w._shapeparts(parts=[s.points], shapeType=s.shapeType)
    w.poly([[[-104, 24], [-104, 25], [-103, 25], [-103, 24], [-104,
    24]]])
    w.record("STANLEY", "TD", "091022/1500", "27", "21", "48", "ep")
```

This is how we do the changes in the original shapefile. Now, let's see how we can add new fields in the shapefile.

Adding fields

A very common operation on shapefiles is to add additional fields to them. This operation is easy but there's one important element to remember. When you add a field, you must also loop through the records and either create an empty cell or add a value for that column. As an example, let's add a reference latitude and longitude column to the UTM version of the New York City museums shapefile:

1. First, we'll open the shapefile and create a new `Writer` object:

    ```
    import shapefile
    r = shapefile.Reader('NYC_MUSEUMS_UTM')
    with shapefile.Writer("test", r.shapeType) as w:
    ```

2. Next, we'll add the fields as float types with a length of 8 for the entire field and a maximum precision of 5 decimal places:

    ```
    w.fields = list(r.fields)
    w.field('LAT','F',8,5)
    w.field('LON','F',8,5)
    ```

3. Next, we'll open the geographic version of the shapefile and grab the coordinates from each record. We'll add these to the corresponding attribute record in the UTM version's `.dbf`:

    ```
    for i in range(len(r.shapes())):
        lon, lat = r.shape(i).points[0]
        w.point(lon, lat)
        w.record(*list(r.record(i)), lat, lon)
    ```

In the next subsection, we'll see how we can merge multiple shapefiles.

Merging shapefiles

Aggregating multiple related shapefiles of the same type into one larger shapefile is another very useful technique. You might be working as part of a team that divides up an area of interest and then assembles the data at the end of the day. Or, you might aggregate data from a series of sensors out in the field, such as weather stations.

For this example, we'll use a set of building footprints for a county that is maintained separately in four different quadrants (northwest, northeast, southwest, and southeast). You can download these shapefiles as a single ZIP file at http://git.io/vLbUE.

When you unzip these files, you'll see they are named by quadrant. The following script uses PyShp to merge them into a single shapefile:

```
import glob
import shapefile
files = glob.glob('footprints_*shp')
with shapefile.Writer("Merged") as w:
    r = None
    for f in files:
        r = shapefile.Reader(f)
        if not w.fields:
            w.fields = list(r.fields)
        for rec in r.records():
            w.record(*list(rec))
        for s in r.shapes():
            w._shapeparts(parts=[s.points], shapeType=s.shapeType)
```

As you can see, merging a set of shapefiles is very straightforward. However, we didn't do any sanity checks to make sure the shapefiles were all of the same type, which you might want to do if this script was used for a repeated automated process, instead of just a quick one-off process.

Another note about this example is how we invoked the `Writer` object. In the other examples, we used a numeric code to define a shape type. You can define that number directly (for example, 1 for point shapefiles) or call one of the PyShp constants. The constants are the type of shapefile in all caps. For example, a polygon is as follows:

```
shapefile.POLYGON
```

In this case, the value of that constant is 5. When copying data from a `Reader` to a `Writer` object, you'll notice the shape type definition is simply referenced, as shown in this example:

```
r = shapefile.Reader('myShape')
w = shapefile.Writer("myShape", r.shapeType)
```

This method makes your script more robust as the script has one less variable that needs to be changed if you later change the script or the dataset. In the merging example, we don't have the benefit of having a `Reader` object available when we invoke `Writer`.

We could open the first shapefile in the list and check its type, but that would add several more lines of code. An easier way is just to omit the shape type. If the `Writer` shape type isn't saved, PyShp will ignore it until you save the shapefile. At that time, it will check the individual header of a geometry record and determine it from that.

While you can use this method in special cases, it's better to define the shape type explicitly when you can, for clarity, and just to be safe to prevent any outlier case errors. The following illustration is a sample of this dataset so that you get a better idea of what the data looks like, as we will be using it more next:

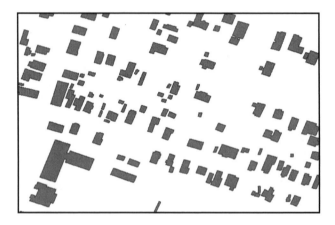

Now, let's see how to do this with the `.dbfpy` files.

Merging shapefiles with dbfpy

The `.dbf` portion of PyShp can occasionally run into issues with `.dbf` files that are produced by certain software. Fortunately, PyShp allows you to manipulate the different shapefile types separately. There's a more robust `.dbf` library, named `dbfpy3`, which we discussed in Chapter 4, *Geospatial Python Toolbox*. You can use PyShp to handle the `.shp` and `.shx` files, while `.dbfpy` handles more complex `.dbf` files. You can download the module here: https://github.com/GeospatialPython/dbfpy3/archive/master.zip.

This approach takes more code but it will often succeed where PyShp alone fails with .dbf issues. This example uses the same shapefiles from the previous example. In the following example, we'll merge a shapefile using only its attributes:

1. First, we import the libraries we need, get a list of shapefiles using the glob module, and create a shapefile `Writer` object using PyShp:

```
import glob
import shapefile
from dbfpy3 import dbf
shp_files = glob.glob('footprints_*.shp')
w = shapefile.Writer(shp="merged.shp", shx="merged.shx")
```

2. Now, we're going to open only the .shp files and copy the geometries to the writer. We'll circle back and get the attributes using the dbypy3 module later to demonstrate working with shapefile components separately:

```
# Loop through ONLY the shp files and copy their shapes
# to a Writer object. We avoid opening the dbf files
# to prevent any field-parsing errors.
for f in shp_files:
    print("Shp: {}".format(f))
    r = shapefile.Reader(f)
    r = shapefile.Reader(shp=shpf)
    for s in r.shapes():
        w.poly([s.points])
    print("Num. shapes: {}".format(len(w.shapes())))
```

3. Once all of the geometry has been copied over to the writer, we can save the .shp file and have PyShp create an index file for the geometry:

```
# Save only the shp and shx index file to the new
# merged shapefile.
w.close()
```

4. Next, we can get a list of .dbf files using the glob module:

```
# Now we come back with dbfpy and merge the dbf files
dbf\_files = glob.glob('\*.dbf')
```

5. Next, we'll use the first .dbf file in the list as a template to get the field data and use it to set the properties of the shapefile writer:

```
# Use the first dbf file as a template
template = dbf\_files.pop(0)
merged\_dbf\_name = 'merged.dbf'
# Copy the entire template dbf file to the merged file
```

```
merged\_dbf = open(merged\_dbf\_name, 'wb')
temp = open(template, 'rb')
merged\_dbf.write(temp.read())
merged\_dbf.close()
temp.close()
```

6. Then, we simply loop through the `.dbf` files and copy the records to `Writer`:

```
# Now read each record from the remaining dbf files
# and use the contents to create a new record in
# the merged dbf file.
db = dbf.Dbf(merged\_dbf\_name)
for f in dbf\_files:
    print('Dbf: {}'.format(f))
    dba = dbf.Dbf(f)
    for rec in dba:
        db\_rec = db.newRecord()
        for k, v in list(rec.asDict().items()):
            db\_rec[k] = v
        db\_rec.store()
db.close()
```

Now that we know how to merge shapefiles, let's check out how to split them.

Splitting shapefiles

Sometimes, you may also need to split larger shapefiles to make it easier for you to focus on a subset of interest. This splitting, or subsetting, can be done spatially or by attributes, depending on which aspect of the data is of interest.

Subsetting spatially

One way to extract part of a dataset is to use spatial attributes such as size. In the following example, we'll subset the southeast quadrant file we merged. We'll filter the building footprint polygons by area and export any buildings with a 100 square meters or less (about 1,000 square feet) profile to a new shapefile. We'll use the `footpints_se` shapefile for this.

PyShp has a signed area method that accepts a list of coordinates and returns either a positive or negative area. We'll use the `utm` module to convert the coordinates into meters. Normally, the positive or negative area denotes whether the point order of the polygon is clockwise or counterclockwise, respectively. But point order doesn't matter here, so we'll use the absolute value using the `abs()` function, as shown here, when we get the area value:

```
import shapefile
import utm
r = shapefile.Reader('footprints\_se')
w = shapefile.Writer(r.shapeType)
w.fields = list(r.fields)
for sr in r.shapeRecords():
    utmPoints = []
    for p in sr.shape.points:
        x,y,band,zone = utm.from_latlon(p[1],p[0])
        utmPoints.append([x,y])
    area = abs(shapefile.signed_area(utmPoints))
    if area <= 100:
        w._shapes.append(sr.shape)
        w.records.append(sr.record)
w.save('footprints\_185')
```

Let's see the difference in the number of records between the original and the subset shapefile:

```
r = shapefile.Reader('footprints\_se')
subset = shapefile.Reader('footprints\_185')
print(r.numRecords)
# 26447
print(subset.numRecords)
# 13331
```

We now have some substantial building blocks for geospatial analysis with vector data, as well as attributes.

Performing selections

The previous subsetting example is one way to select data. There are many other ways to subset data for further analysis. In this section, we'll examine selecting subsets of data that are critical for efficient data processing to reduce the size of a large dataset down to just our area of interest for a given dataset.

The point-in-polygon formula

We briefly discussed the point-in-polygon formula in Chapter 1, *Learning about Geospatial Analysis with Python*, as a common type of geospatial operation. You'll find it is one of the most useful formulas out there. The formula is relatively straightforward.

The following function performs this check using the **Ray Casting** method. This method draws a line from the test point all of the way through the polygon and counts the number of times it crosses the polygon boundary. If the count is even, the point is outside the polygon. If it is odd, then it's inside. This particular implementation also checks to see whether the point is on the edge of the polygon:

```python
def point_in_poly(x,y,poly):
    # check if point is a vertex
    if (x,y) in poly: return True
    # check if point is on a boundary
    for i in range(len(poly)):
        p1 = None
        p2 = None
        if i==0:
            p1 = poly[0]
            p2 = poly[1]
        else:
            p1 = poly[i-1]
            p2 = poly[i]
        if p1[1] == p2[1] and p1[1] == y and x min(p1[0], \
            p2[0]) and x < max(p1[0], p2[0]):
            return True
    n = len(poly)
    inside = False
    p1x,p1y = poly[0]
    for i in range(n+1):
        p2x,p2y = poly[i % n]
        if y min(p1y,p2y):
            if y <= max(p1y,p2y):
                if x <= max(p1x,p2x):
                    if p1y != p2y:
                        xints = (y-p1y)*(p2x-p1x)/(p2y-p1y)+p1x
                    if p1x == p2x or x <= xints:
                        inside = not inside
        p1x,p1y = p2x,p2y
    if inside: return True
    return False
```

Now, let's use the `point_in_poly()` function to test a point in Chile:

```
# Test a point for inclusion
myPolygon = [(-70.593016,-33.416032), (-70.589604,-33.415370),
(-70.589046,-33.417340), (-70.592351,-33.417949),
(-70.593016,-33.416032)]
# Point to test
lon = -70.592000
lat = -33.416000
print(point_in_poly(lon, lat, myPolygon))
# True
```

This shows that the point is inside. Let's also verify that edge points will be detected:

```
# test an edge point
lon = -70.593016
lat = -33.416032
print(point_in_poly(lon, lat, myPolygon))
# True
```

You'll find new uses for this function all the time. It's definitely one to keep in your toolbox.

Bounding box selections

A bounding box is the smallest rectangle that can completely contain a feature. We can use it as an efficient way to subset one or more individual features from a larger dataset. We'll look at one more example of using a simple bounding box to isolate a complex set of features and save it in a new shapefile. In this example, we'll subset the roads on the island of Puerto Rico from the mainland US Major Roads shapefile. You can download the shapefile here: `https://github.com/GeospatialPython/Learn/raw/master/roads.zip`.

Floating-point coordinate comparisons can be expensive, but because we are using a box and not an irregular polygon, this code is efficient enough for most operations:

```
import shapefile
r = shapefile.Reader('roadtrl020')
w = shapefile.Writer(r.shapeType)
w.fields = list(r.fields)
xmin = -67.5
xmax = -65.0
ymin = 17.8
ymax = 18.6
for road in r.iterShapeRecords():
    geom = road.shape
    rec = road.record
    sxmin, symin, sxmax, symax = geom.bbox
```

```
        if sxmin < xmin: continue
        elif sxmax xmax: continue
        elif symin < ymin: continue
        elif symax ymax: continue
        w._shapes.append(geom)
        w.records.append(rec)
    w.save('Puerto_Rico_Roads')
```

Now that we've used geometry to select features, let's do it another way by using attributes.

Attribute selections

We've now seen two different ways of subsetting a larger dataset, resulting in a smaller one based on spatial relationships. But we can also select data using the attribute fields. So, let's examine a quick way to subset vector data using the attribute table. In this example, we'll use a polygon shapefile that has densely populated urban areas within Mississippi. You can download this zipped shapefile from http://git.io/vLbU9.

This script is really quite simple. It creates the Reader and Writer objects, copies the .dbf fields, loops through the records for matching attributes, and then adds them to Writer. We'll select urban areas with a population of less than 5000:

```
import shapefile
# Create a reader instance
r = shapefile.Reader('MS_UrbanAnC10')
# Create a writer instance
w = shapefile.Writer(r.shapeType)
# Copy the fields to the writer
w.fields = list(r.fields)
# Grab the geometry and records from all features
# with the correct population
selection = []
for rec in enumerate(r.records()):
    if rec[1][14] < 5000:
        selection.append(rec)
# Add the geometry and records to the writer
for rec in selection:
    w._shapes.append(r.shape(rec[0]))
    w.records.append(rec[1])
# Save the new shapefile
w.save('MS_Urban_Subset')
```

Attribute selections are typically fast. Spatial selections are computationally expensive because of floating-point calculations. Whenever possible, make sure you are unable to use attribute selection to subset first. The following illustration shows the starting shapefile containing all urban areas on the left with a state boundary, and the urban areas with less than 5,000 people on the right, after the previous attribute selection:

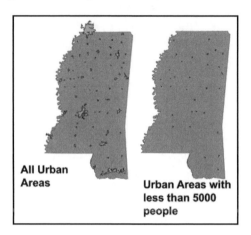

All Urban Areas

Urban Areas with less than 5000 people

Let's see what that same example looks like using `fiona`, which takes advantage of the OGR library. We'll use nested `with` statements to reduce the amount of code needed to properly open and close the files:

```
import fiona
with fiona.open('MS_UrbanAnC10.shp') as sf:
    filtered = filter(lambda f: f['properties']['POP'] < 5000, sf)
    # Shapefile file format driver
    drv = sf.driver
    # Coordinate Reference System
    crs = sf.crs
    # Dbf schema
    schm = sf.schema
    subset = 'MS_Urban_Fiona_Subset.shp'
    with fiona.open(subset, 'w',
        driver=drv,
        crs=crs,
        schema=schm) as w:
            for rec in filtered:
                w.write(rec)
```

Now, we know how to combine discrete datasets as well as split larger datasets apart. What else can we do? We can aggregate features within a dataset.

Aggregating geometry

GIS vector datasets are typically composed of point, line, or polygon features. One of the principles of GIS is that things that are closer together geographically are more related than things that are further apart. When you have a set of related features, often, it's too much detail for the analysis you're trying to accomplish. It can be useful to generalize them to speed up processing or simplify a map. This type of operation is called **aggregation**. A common example of aggregation is to combine a set of local political boundaries into a larger political boundary such as counties into a state or states into a country or countries into continents.

In this example, we'll do just that. We'll convert a dataset comprising all of the counties in the US state of Mississippi into a single polygon representing the entire state. The Python Shapely library is perfect for this kind of operation; however, it can only manipulate geometry and doesn't read or write data files. To read and write data files, we'll use the Fiona library. If you don't have Shapely or Fiona installed, use `pip` to install them. You can download the counties dataset here: `https://git.io/fjt3b`.

The following illustration shows what the counties dataset looks like:

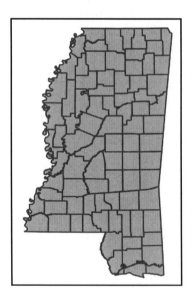

The following steps will show you how to merge the individual county polygons into a single polygon:

1. In the following code, we import the libraries we need, including the different portions of the `shapely` library.
2. Then, we'll open the counties GeoJSON file.
3. Next, we'll copy the schema of the source file, which defines all of the metadata for the dataset.
4. Then, we need to alter that metadata copy to change the attributes in order to define a single attribute for the state name. We also need to alter the geometry type from *MultiPolygon* to *Polygon*.
5. Then, we'll open our output dataset GeoJSON file named `combined.geojson`.
6. Next, we'll extract all of the polygons and attributes and combine all of the polygons into one.
7. Finally, we'll write the combined polygon out with the new attribute.
8. We'll import our libraries, including `OrderDict`, so that we can maintain control of the shapefile attributes:

```
# Used OrderedDict to control the order
# of data attributes
from collections import OrderedDict
# Import the shapely geometry classes and methods.
# The "mapping" method returns a GeoJSON representation
# of a geometry.
from shapely.geometry import shape, mapping, Polygon
# Import the shapely union function which combines
# geometries
from shapely.ops import unary_union
# Import Fiona to read and write datasets
import fiona
```

9. We open our GeoJSON file and copy the metadata:

```
# Open the counties dataset
with fiona.open('ms_counties.geojson') as src:
    # copy the metadata
    schema = src.meta.copy()
    # Create a new field type for our
    # state dataset
    fields = {"State": "str:80"}
```

10. Then, we create our new field:

```
# Create a new property for our dataset
# using the new field
prop = OrderedDict([("State", "Mississippi")])
# Change the metadata geometry type to Polygon
schema['geometry'] = 'Polygon'
schema['schema']['geometry'] = 'Polygon'
```

11. Now, we can add the new field to the metadata:

```
# Add the new field
schema['properties'] = fields
schema['schema']['properties'] = fields
```

12. Next, we can open the combined GeoJSON file and write out our results:

```
# Open the output GeoJSON dataset
with fiona.open('combined.geojson', 'w', **schema) as dst:
    # Extract the properties and geometry
    # from the counties dataset
    props, geom = zip(*[(f['properties'],shape(f['geometry'])) for
    f in src])
    # Write the new state dataset out while
    # combining the polygons into a
    # single polygon and add the new property
    dst.write({'geometry': mapping(\
    Polygon(unary_union(geom).exterior)),
    'properties': prop})
```

The output dataset will look similar to the following illustration:

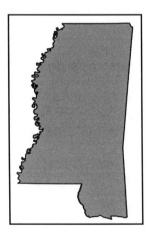

Now that we know all about reading, editing, and writing GIS data, we can begin visualizing it in the upcoming sections.

Creating images for visualization

Now, we're moving from calculations and data editing to something we can see! We'll begin by creating different types of maps. In Chapter 1, *Learning about Geospatial Analysis with Python*, we visualized our SimpleGIS program using the Tkinter module that's included with Python. In Chapter 4, *Geospatial Python Toolbox*, we examined a few other methods for creating images. Now, we'll examine these tools in more depth by creating two specific types of thematic maps. The first is a dot density map and the second is a choropleth map.

First, let's start with the dot density map.

Dot density calculations

A dot density map shows concentrations of subjects within a given area. If an area is divided up into polygons containing statistical information, you can model that information using randomly distributed dots within that area using a fixed ratio across the dataset. This type of map is commonly used for population density maps.

The cat map in Chapter 1, *Learning about Geospatial Analysis with Python*, is a dot density map. Let's create a dot density map from scratch using pure Python. Pure Python allows you to work with much more lightweight libraries that are generally easier to install and are more portable. For this example, we'll use a US Census Bureau Tract shapefile along the US Gulf Coast, which contains population data. We'll also use the point-in-polygon algorithm to ensure the randomly distributed points are within the proper census tract. Finally, we'll use the PNGCanvas module to write out our image.

The PNGCanvas module is excellent and fast. However, it doesn't have the ability to fill in polygons beyond simple rectangles. You can implement a fill algorithm but it is very slow in pure Python. However, for a quick outline and point plot, it does a great job.

You'll also see the `world2screen()` method, which is similar to the coordinates-to-mapping algorithm we used in SimpleGIS in Chapter 1, *Learning about Geospatial Analysis with Python*. In this example, we'll read in a shapefile and write it back out as an image:

1. First, we import the libraries we need, including `pngcanvas`, to draw a map image:

```
import shapefile
import random
import pngcanvas
```

2. Next, we define our point-in-polygon function, which we've used before. In this example, we'll use it to randomly distribute population values within a location:

```
def point_in_poly(x,y,poly):
    '''Boolean: is a point inside a polygon?'''
    # check if point is a vertex
    if (x,y) in poly: return True
    # check if point is on a boundary
    for i in range(len(poly)):
        p1 = None
        p2 = None
        if i==0:
            p1 = poly[0]
            p2 = poly[1]
        else:
            p1 = poly[i-1]
            p2 = poly[i]
        if p1[1] == p2[1] and p1[1] == y and \
        x min(p1[0], p2[0]) and x < max(p1[0], p2[0]):
            return True
    n = len(poly)
    inside = False
    p1x,p1y = poly[0]
    for i in range(n+1):
        p2x,p2y = poly[i % n]
        if y min(p1y,p2y):
            if y <= max(p1y,p2y):
                if x <= max(p1x,p2x):
                    if p1y != p2y:
                        xints = (y-p1y)*(p2x-p1x)/(p2y-p1y)+p1x
                    if p1x == p2x or x <= xints:
                        inside = not inside
        p1x,p1y = p2x,p2y
    if inside: return True
    else: return False
```

3. Now, we need a function to scale our geospatial coordinates to the map image:

```
def world2screen(bbox, w, h, x, y):
    '''convert geospatial coordinates to pixels'''
    minx,miny,maxx,maxy = bbox
    xdist = maxx - minx
    ydist = maxy - miny
    xratio = w/xdist
    yratio = h/ydist
    px = int(w - ((maxx - x) * xratio))
    py = int((maxy - y) * yratio)
    return (px,py)
```

4. Next, we read in the shapefile and set the size of our output map image:

```
# Open the census shapefile
inShp = shapefile.Reader('GIS_CensusTract_poly')
# Set the output image size
iwidth = 600
iheight = 400
```

5. Next, we need to determine the index of the population field so that we can get the population count for each area:

```
# Get the index of the population field
pop_index = None
dots = []
for i,f in enumerate(inShp.fields):
    if f[0] == 'POPULAT11':
        # Account for deletion flag
        pop_index = i-1
```

6. Then, we calculate the population density value. We want to create a dot on the map for every 100 people:

```
# Calculate the density and plot points
for sr in inShp.shapeRecords():
    population = sr.record[pop_index]
    # Density ratio - 1 dot per 100 people
    density = population / 100
    found = 0
```

7. We will loop through each polygon and randomly distribute the points to create a density map:

```
# Randomly distribute points until we
# have the correct density
while found < density:
    minx, miny, maxx, maxy = sr.shape.bbox
    x = random.uniform(minx,maxx)
    y = random.uniform(miny,maxy)
    if point_in_poly(x,y,sr.shape.points):
        dots.append((x,y))
        found += 1
```

8. We're now ready to create our output image:

```
# Set up the PNG output image
c = pngcanvas.PNGCanvas(iwidth,iheight)
# Draw the red dots
c.color = (255,0,0,0xff)
for d in dots:
    # We use the *d notation to exand the (x,y) tuple
    x,y = world2screen(inShp.bbox, iwidth, iheight, *d)
    c.filled_rectangle(x-1,y-1,x+1,y+1)
```

9. Our dots have been created. Now, we need to create the outlines of the census tract:

```
# Draw the census tracts
c.color = (0,0,0,0xff)
for s in inShp.iterShapes():
    pixels = []
    for p in s.points:
        pixel = world2screen(inShp.bbox, iwidth, iheight, *p)
        pixels.append(pixel)
    c.polyline(pixels)
```

10. Finally, we'll save the output image:

```
# Save the image
with open('DotDensity.png','wb') as img:
    img.write(c.dump())
```

This script outputs an outline of the census tract, along with the density dots, to show population concentration very effectively:

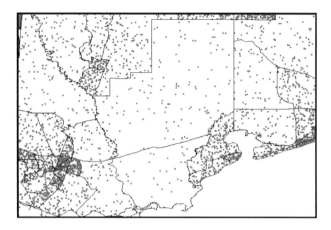

Now, let's check out the second type of map: choropleth maps.

Choropleth maps

A choropleth uses shading, coloring, or symbols to show an average value or quantity within an area. They make it easy for us to visualize large amounts of data as a summary. This method is useful if related data spans multiple polygons. For example, in a worldwide population density map by country, many countries have disconnected polygons (for example, Hawaii is an island state of the US).

In this example, we'll use the **Python Imaging Library** (**PIL**) we discussed in Chapter 3, *The Geospatial Technology Landscape*. PIL is not purely Python but is designed specifically for Python. We'll recreate our previous dot density example as a choropleth map. We'll calculate a density ratio for each census tract based on the number of people (population) per square kilometer and use that value to adjust the color. Dark is more densely populated while lighter is less. Follow these steps:

1. First, we will import our libraries:

```
import math
import shapefile
try:
    import Image
    import ImageDraw
except:
    from PIL import Image, ImageDraw
```

2. Then, we'll need our geospatial coordinates to image coordinates conversion function:

```
def world2screen(bbox, w, h, x, y):
    '''convert geospatial coordinates to pixels'''
    minx,miny,maxx,maxy = bbox
    xdist = maxx - minx
    ydist = maxy - miny
    xratio = w/xdist
    yratio = h/ydist
    px = int(w - ((maxx - x) * xratio))
    py = int((maxy - y) * yratio)
    return (px,py)
```

3. Now, we open our shapefile and set our output image size:

```
# Open our shapefile
inShp = shapefile.Reader('GIS_CensusTract_poly')
iwidth = 600
iheight = 400
```

4. We then set up PIL to draw our map image:

```
# PIL Image
img = Image.new('RGB', (iwidth,iheight), (255,255,255))
# PIL Draw module for polygon fills
draw = ImageDraw.Draw(img)
```

5. Just like our previous example, we need to get the index of the population field:

```
# Get the population AND area index
pop_index = None
area_index = None
# Shade the census tracts
for i,f in enumerate(inShp.fields):
    if f[0] == 'POPULAT11':
        # Account for deletion flag
        pop_index = i-1
    elif f[0] == 'AREASQKM':
        area_index = i-1
```

6. Now, we can draw the polygons, shade them according to population density, and save the image:

```
# Draw the polygons
for sr in inShp.shapeRecords():
    density = sr.record[pop_index]/sr.record[area_index]
    # The 'weight' is a scaled value to adjust the color
```

```
# intensity based on population
weight = min(math.sqrt(density/80.0), 1.0) * 50
R = int(205 - weight)
G = int(215 - weight)
B = int(245 - weight)
pixels = []
for x,y in sr.shape.points:
    (px,py) = world2screen(inShp.bbox, iwidth, iheight, x, y)
    pixels.append((px,py))
    draw.polygon(pixels, outline=(255,255,255), fill=(R,G,B))
img.save('choropleth.png')
```

This script produces the following diagram with the relative density of tracks. You can adjust the color using the R, G, and B variables:

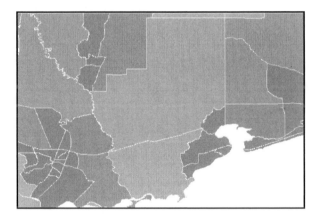

Now that we can show statistical data from shapefiles, we can look at a statistical data source that is even more common than shapefiles: spreadsheets.

Using spreadsheets

Spreadsheets such as Microsoft Office Excel and Open Office Calc are inexpensive (even free), ubiquitous, easy to use, and great for recording structured data. For these reasons, spreadsheets are widely used to collect data for entry into a GIS format. As an analyst, you will find yourself working with spreadsheets frequently.

In the previous chapters, we discussed the CSV format, which is a text file with the same basic rows and columns data structure as a spreadsheet. For CSV files, you use Python's built-in `csv` module. But most of the time, people don't bother exporting a true spreadsheet to a generic CSV file. That's where the pure Python `xlrd` module comes into play. The name `xlrd` is short for **Excel Reader** and is available from PyPI. There is also an accompanying module, the `xlwt` (Excel Writer) module, for writing spreadsheets. These two modules make reading and writing Excel spreadsheets a snap. Combine it with PyShp and you can move back and forth between spreadsheets and shapefiles with ease. This example demonstrates converting a spreadsheet into a shapefile. We'll use a spreadsheet version of the New York City museum point data available at `https://git.io/Jemi9`.

The spreadsheet contains the attribute data, followed by an *x* column with the longitude and a *y* column with the latitude. To export it to a shapefile, we'll execute the following steps:

1. Open the spreadsheet.
2. Create a shapefile `Writer` object.
3. Capture the first row of the spreadsheet as the `dbf` columns.
4. Loop through each row of the spreadsheet and copy the attributes to `dbf`.
5. Create a point from the *x* and *y* spreadsheet columns.

The script is as follows:

```
import xlrd
import shapefile
# Open the spreadsheet reader
xls = xlrd.open_workbook('NYC_MUSEUMS_GEO.xls')
sheet = xls.sheet_by_index(0)
# Open the shapefile writer
w = shapefile.Writer(shapefile.POINT)
# Move data from spreadsheet to shapefile
for i in range(sheet.ncols):
    # Read the first header row
    w.field(str(sheet.cell(0,i).value), 'C', 40)
for i in range(1, sheet.nrows):
    values = []
    for j in range(sheet.ncols):
        values.append(sheet.cell(i,j).value)
    w.record(*values)
    # Pull latitude/longitude from the last two columns
    w.point(float(values[-2]),float(values[-1]))
w.save('NYC_MUSEUMS_XLS2SHP')
```

Converting a shapefile into a spreadsheet is a much less common operation, though not difficult. To convert a shapefile into a spreadsheet, you need to make sure you have an *x* and *y* column by using the *Adding fields* example from the *Editing shapefiles* section in this chapter. You would loop through the shapes and add the *x*, *y* values to those columns. Then, you would read the field names and column values from dbf into an xlwt spreadsheet object or a CSV file using the csv module. The coordinate columns are labeled in the following screenshot:

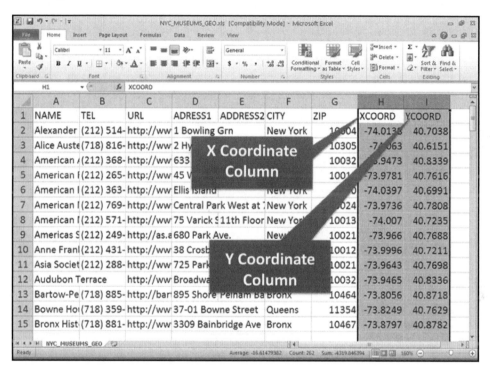

In the next section, we'll use a spreadsheet as an input data source.

Creating heat maps

A heat map is used to show the geographic clustering of data using a raster image that shows density. The clustering can also be weighed by using a field in the data to not only show geographic density but also an intensity factor. In this example, we'll use bear sighting data contained in the CSV dataset, which stores the data as points to create a heat map of the frequency of bear sightings in different areas of Mississippi. This dataset is so simple that's we're going to treat the CSV file as a text file, which is one of the nice features of a CSV file.

You can download the dataset here: `https://git.io/fjtGL`.

The output is going to be a simple HTML web map that you can open in any web browser. The web map will be based on the excellent Leaflet JavaScript library. On top of that, we'll use the Python Folium library, which makes it easy for us to create Leaflet web maps, in order to generate the HTML page:

```
import os
import folium
from folium.plugins import HeatMap
f = open('bear_sightings.csv', 'r')
lines = f.readlines()
lines.pop(0)
data = []
bears = [list(map(float, l.strip().split(','))) for l in lines]
m = folium.Map([32.75, -89.52], tiles='stamentonerbackground',
zoom_start=7, max_zoom=7, min_zoom=7)
HeatMap(bears, max_zoom=16, radius=22, min_opacity=1, blur=30).add_to(m)
m.save('heatmap.html')
```

This script will create a file called `heatmap.html`. Open it in any web browser to see a similar image:

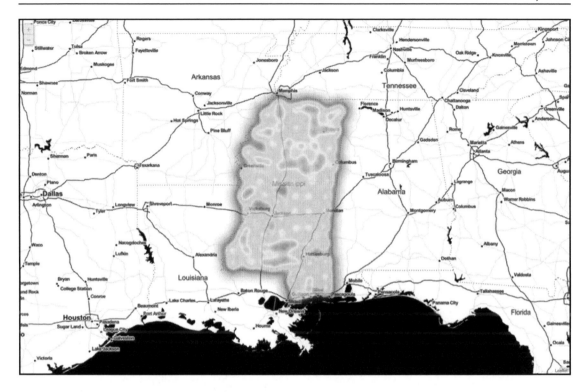

Next, we'll learn how to use data generated by a GPS to collect field data such as the information in the preceding heatmap.

Using GPS data

The most common type of GPS data these days is the Garmin GPX format. We covered this XML format in `Chapter 4`, *Geospatial Python Toolbox*, which has become an unofficial industry standard. Because it is an XML format, all of the well-documented rules of XML apply to it. However, there is another type of GPS data that pre-dates XML and GPX, called the **National Marine Electronics Association (NMEA)**. This data is ASCII text sentences that are designed to be streamed.

You occasionally bump into this format from time to time because even though it is older and esoteric, it is still very much alive and well, especially for communicating ship locations via the **Automated Identification System (AIS)**, which tracks ships globally. But as usual, you have a good option in pure Python. The pynmea module is available on PyPI. The following code is a small sample of NMEA sentences:

```
$GPRMC,012417.859,V,1856.599,N,15145.602,W,12.0,7.27,020713,,E\*4F
$GPGGA,012418.859,1856.599,N,15145.602,W,0,00,,,M,,M,,\*54
$GPGLL,1856.599,N,15145.602,W,012419.859,V\*35
$GPVTG,7.27,T,,M,12.0,N,22.3,K\*52
$GPRMC,012421.859,V,6337.596,N,12330.817,W,66.2,23.41,020713,,E\*74
```

To install the pynmea module from PyPI and download the complete sample file, you can view the following URL: http://git.io/vLbTv. Then, you can run the following sample, which will parse the NMEA sentences into objects. The NMEA sentences contain a wealth of information:

```
from pynmea.streamer import NMEAStream
nmeaFile = open('nmea.txt')
nmea_stream = NMEAStream(stream_obj=nmeaFile)
next_data = nmea_stream.get_objects()
nmea_objects = []
while next_data:
    nmea_objects += next_data
    next_data = nmea_stream.get_objects()
# The NMEA stream is parsed!
# Let's loop through the
# Python object types:
for nmea_ob in nmea_objects:
    if hasattr(nmea_ob, 'lat'):
        print('Lat/Lon: (%s, %s)' % (nmea_ob.lat, nmea_ob.lon))
```

The latitudes and longitudes are stored in a format called **degrees decimal minutes**. For example, this random coordinate, 4533.35, is 45 degrees and 33.35 minutes. 0.35 of a minute is exactly 21 seconds. In another example, 16708.033 is 167 degrees and 8.033 minutes. 0.033 of a minute is approximately 2 seconds. You can find more information about the NMEA format at http://aprs.gids.nl/nmea/.

GPS data is an important location data source, but there's another way we can describe a point on the Earth using a street address. The method for locating a street address on the Earth is called geocoding.

Geocoding

Geocoding is the process of converting a street address into latitude and longitude. This operation is critical to in-vehicle navigation systems and online driving direction websites. Python has two excellent geocoder libraries available named `geocoder` and `geopy`. Both take advantage of online geocoding services to allow you to geocode addresses programmatically. The geopy library even lets you reverse geocode to match a latitude and longitude to the nearest address:

1. First, let's do a quick example with the `geocoder` library, which defaults to using Google Maps as its engine:

```
import geocoder
g = geocoder.google('1403 Washington Ave, New Orleans, LA 70130')
print(g.geojson)
# {'type': 'Feature', 'geometry': {'type': 'Point', 'coordinates':
[-90.08421849999999, 29.9287839]},
'bbox': {'northeast': [29.9301328802915, -90.0828695197085],
'southwest': [29.9274349197085, -90.0855674802915]},
'properties': {'quality': 'street_address', 'lat': 29.9287839,
'city': 'New Orleans',
'provider': 'google', 'geometry': {'type': 'Point', 'coordinates':
[-90.08421849999999, 29.9287839]},
'lng': -90.08421849999999, 'method': 'geocode', 'encoding':
'utf-8', 'confidence': 9, 'address': '1403 Washington Ave,
New Orleans, LA 70130, USA', 'ok': True, 'neighborhood': 'Garden
District', 'county': 'Orleans Parish',
'accuracy': 'ROOFTOP', 'street': 'Washington Ave', 'location':
'1403 Washington Ave, New Orleans, LA 70130',
'bbox': {'northeast': [29.9301328802915, -90.0828695197085],
'southwest': [29.9274349197085, -90.0855674802915]},
'status': 'OK', 'country': 'US', 'state': 'LA', 'housenumber':
'1403', 'postal': '70130'}}
print(g.wkt)
# 'POINT(-90.08421849999999 29.9287839)'
```

Here, we print the GeoJSON record for that address, which contains all known information in Google's database. Then, we print out the returned latitude and longitude as a WKT string, which could be used as input to other operations such as checking whether the address is inside of a flood plain polygon. The documentation for this library also shows you how to switch to other online geocoding services such as Bing or Yahoo. Some of these services require an API key and may have request limits.

2. Now, let's look at the `geopy` library. In this example, we'll geocode against the `OpenStreetMap` database. Once we match the address to a location, we'll turn around and reverse geocode it:

```
from geopy.geocoders import Nominatim
g = Nominatim()
location = g.geocode('88360 Diamondhead Dr E, Diamondhead, MS 39525')
rev = g.reverse('{},{}'.format(location.latitude, location.longitude))
print(rev)
# NVision Solutions Inc., 88360, Diamondhead Drive East, Diamondhead,
Hancock County, Mississippi, 39520,
# United States of America
print(location.raw)
# {'class': 'office', 'type': 'yes', 'lat': '30.3961962', 'licence': 'Data
© OpenStreetMap contributors,
# ODbL 1.0. http://www.openstreetmap.org/copyright', 'display\_name':
'NVision Solutions Inc.,
# 88360, Diamondhead Drive East, Diamondhead, Hancock County, Mississippi,
39520, United States of America',
# 'lon': '-89.3462139', 'boundingbox': ['30.3961462', '30.3962462',
'-89.3462639', '-89.3461639'],
# 'osm\_id': '2470309304', 'osm\_type': 'node', 'place\_id': '25470846',
'importance': 0.421}
```

Now that we know of several different ways to geocode, let's look at speeding up the process. If you have thousands of addresses to geocode, it can take a while. Using multiprocessing, you can reduce a geocoding process that might take days into a few hours.

Multiprocessing

Geospatial datasets are very large. Processing them can take time, which can be hours or sometimes even days. But there's a way you can speed processing up for certain operations. Python's built-in multiprocessing module can spawn multiple processes on your computer to take advantage of all of the available processors.

One operation that works really well with the multiprocessing module is geocoding. In this example, we'll geocode a list of cities and split that processing across all of the processors on your machine. We'll use the same geocoding technique as before, but this time, we'll add the multiprocessing module to increase the potential for greater speed and scalability. The following code will geocode a list of cities simultaneously across multiple processors:

1. First, we import the modules we need:

```
# Import our geocoding module
from geopy.geocoders import Nominatim
# Import the multiprocessing module
import multiprocessing as mp
```

2. Next, we create our geocoder object:

```
# Create our geocoder
g = Nominatim()
```

3. Now, we need a function to geocode and individual address:

```
# Create a function to geocode an individual address
def gcode(address):
    location = g.geocode(address)
    print("Geocoding: {}".format(address))
    return location
```

4. Next, we create our list of cities to process:

```
# Our list of cities to process
cities = ["New Orleans, LA", "Biloxi, MS", "Memphis, TN",
"Atlanta, GA", "Little Rock, AR", "Destin, FL"]
```

5. Then, we set up our processor pool based on the number of processors available:

```
# Create our processor pool counting all of the processors
# on the machine.
pool = mp.Pool(processes=mp.cpu_count())
```

6. Next, we map our list of cities to the geocode function though the processor pool:

```
# Map our cities list to the geocoding function
# and allow the processor pool to split it
# across processors
results = pool.map(gcode, cities)
```

7. Then, we can print the results:

```
# Now print the results
print(results)

# [Location(New Orleans, Orleans Parish, Louisiana, USA,
(29.9499323, -90.0701156, 0.0)),
# Location(Biloxi, Harrison County, Mississippi, USA, (30.374673,
-88.8459433348286, 0.0)),
# Location(Memphis, Shelby County, Tennessee, USA, (35.1490215,
-90.0516285, 0.0)),
# Location(Atlanta, Fulton County, Georgia, USA, (33.7490987,
-84.3901849, 0.0)),
# Location(Little Rock, Arkansas, USA, (34.7464809, -92.2895948,
0.0)),
# Location(Destin, Okaloosa County, Florida, USA, (30.3935337,
-86.4957834, 0.0))]
```

This technique can be very powerful, but not every type of processing can be performed this way. The type of processing you use has to support operations that can be broken apart into discrete calculations. But when you can break problems apart, like we did in this example, the results are orders of magnitude faster.

Summary

This chapter covered the critical components of GIS analysis. We examined the challenges of measuring on the curved surface of the Earth using different approaches. We looked at the basics of coordinate conversion and full reprojection using OGR, the utm module with PyShp, and Fiona, which simplifies OGR. We edited shapefiles and performed spatial and attribute selections. We created thematic maps from scratch using only Python. We also imported data from spreadsheets. Then, we parsed GPS data from NMEA streams. Finally, we used geocoding to convert street addresses into locations and back.

As a geospatial analyst, you may be familiar with both GIS and remote sensing, but most analysts specialize in one field or the other. That is why this book approaches the fields in separate chapters – so that we can focus on their differences. As we mentioned in the introduction, the techniques in this chapter are the building blocks for all geospatial analysis and will give you the tools you need so that you can learn about any aspect of this field.

In Chapter 6, *Python and Remote Sensing*, we'll tackle remote sensing. In GIS, we have been able to explore this field using pure Python modules. In remote sensing, we'll become more dependent on bindings to compiled modules written in C due to the sheer size and complexity of the data.

Python and Remote Sensing

6

In this chapter, we will discuss remote sensing. Remote sensing is about gathering a collection of information about the Earth without making physical contact with it. Typically, this means having to use satellite or aerial imagery, **Light Detection and Ranging (LIDAR)**, which measures laser pulses from an aircraft to the Earth, or synthetic aperture radar. Remote sensing can also refer to processing data that's been collected, which is how we'll use the term in this chapter. Remote sensing grows in a more exciting way every day as more satellites are launched and the distribution of data becomes easier. The high availability of satellite and aerial images, as well as interesting new types of sensors launching each year, is changing the role that remote sensing plays in understanding our world.

In remote sensing, we step through each pixel in an image and perform some form of query or mathematical process. An image can be thought of as a large numerical array. In remote sensing, these arrays can be quite large, in the order of tens of megabytes to several gigabytes in size. While Python is fast, only C-based libraries can provide the speed that's needed to loop through arrays at a tolerable speed.

We'll use the **Python Imaging Library (PIL)** for image processing and NumPy, which provides multidimensional array mathematics. While written in C for speed, these libraries are designed for Python and provide a Pythonic API.

In this chapter, we'll cover the following topics:

- Swapping image bands
- Creating image histograms
- Performing a histogram stretch
- Clipping and classifying images
- Extracting features from images
- Change detection

First, we'll start with basic image manipulation and then build on each exercise, all the way to automatic change detection. These techniques will compliment the previous chapters by adding the ability to process satellite data and other remote sensing products to our toolbox.

Technical requirements

- Python 3.6 or higher
- RAM: Minimum 6 GB (Windows), 8 GB (macOS), recommended 8 GB
- Storage: Minimum 7200 RPM SATA with 20 GB of available space; recommended SSD with 40 GB of available space
- Processor: Minimum Intel Core i3 2.5 GHz; recommended Intel Core i5

Swapping image bands

Our eyes can only see colors in the visible spectrum as combinations of **red, green, and blue** (**RGB**). Air and space-borne sensors can collect wavelengths of the energy outside of the visible spectrum. To view this data, we move images representing different wavelengths of light reflectance in and out of the RGB channels to make color images.

These images often end up as bizarre and alien color combinations that can make visual analysis difficult. An example of a typical satellite image is shown in the following Landsat 7 satellite scene near the NASA Stennis Space Center in Mississippi along the Gulf of Mexico, which is a leading center for remote sensing and geospatial analysis in general:

Most of the vegetation appears red and water appears almost black. This image is a type of false-color image, meaning the color of the image is not based on the RGB light. However, we can change the order of the bands or swap out certain bands to create another type of false-color image that looks more like the world we are used to seeing. To do so, you first need to download this image as a ZIP file from here: https://git.io/vqs41.

We installed the GDAL library with Python bindings in Chapter 4, *Geospatial Python Toolbox*, in the *Installing GDAL and NumPy* section. The GDAL library includes a module called gdal_array that loads and saves remotely-sensed images to and from NumPy arrays for easy manipulation. GDAL itself is a data access library and does not provide much in the name of processing. So, in this chapter, we will rely heavily on NumPy to actually change images.

In this example, we'll load the image into a NumPy array using `gdal_array` and then we'll immediately save it back to a new GeoTiff file. However, upon saving, we'll use NumPy's advanced array-slicing feature to change the order of the bands. Images in NumPy are multi-dimensional arrays in the order of band, height, and width. This means that an image with three bands will be an array of length 3, containing an array for the band, height, and width of the image. It's important to note that NumPy references array locations as *y,x (row, column)* instead of the usual *x, y (column, row)* format we work with in spreadsheets and other software. Let's get started:

1. First, we'll import `gdal_array`:

   ```
   from gdal import gdal_array
   ```

2. Next, we'll load an image named `FalseColor.tif` into a `numpy` array:

   ```
   # name of our source image
   src = "FalseColor.tif"
   # load the source image into an array
   arr = gdal_array.LoadFile(src)
   ```

3. Next, we'll reorder the image bands by slicing the array, rearranging the order, and saving it back out:

   ```
   # swap bands 1 and 2 for a natural color image.
   # We will use numpy "advanced slicing" to reorder the bands.
   # Using the source image
   output = gdal_array.SaveArray(arr[[1, 0, 2], :], "swap.tif",
    format="GTiff", prototype=src)
   # Dereference output to avoid corrupted file on some platforms
   output = None
   ```

In the `SaveArray` method, the last argument is called a **prototype**. This argument lets you specify another image for GDAL from which you copy spatial reference information and some other image parameters. Without this argument, we'd end up with an image without georeferencing information, which could not be used in a GIS. In this case, we specify our input image file name because the images are identical, except for the band order. In this method, you can tell that the Python GDAL API is a wrapper around a C library and is not as Pythonic as a Python-designed library. For example, a pure Python library would have written the `SaveArray()` method as `save_array()` to follow Python standards.

The result of this example produces the `swap.tif` image, which is a much more visually appealing image with green vegetation and blue water:

There's only one problem with this image: it's kind of dark and difficult to see. Let's see if we can figure out why in the next section.

Creating histograms

A histogram shows the statistical frequency of data distribution within a dataset. In the case of remote sensing, the dataset is an image. The data distribution is the frequency of pixels in the range of **0** to **255**, which is the range of 8-byte numbers that are used to store image information on computers.

In an RGB image, color is represented as a 3-digit tuple with *(0,0,0, 0, 0)* being black and *(255,255,255)* being white. We can graph the histogram of an image with the frequency of each value along the y-axis and the range of 256 possible pixel values along the x-axis.

Remember in Chapter 1, *Learning about Geospatial Analysis with Python*, in the *Creating the simplest possible Python GIS* section, when we used the Turtle graphics engine included with Python to create a simple GIS? Well, we can also use it to easily graph histograms.

Histograms are usually a one-off product that makes a quick script. Also, histograms are typically displayed as a bar graph with the width of the bars representing the size of grouped data bins. But, in an image, each bin is only one value, so we'll create a line graph. We'll use the histogram function in this example and create a red, green, and blue line for each respective band.

The graphing portion of this example also defaults to scaling the *y*-axis values to the max RGB frequency found in the image. Technically, the *y*-axis represents the maximum frequency, which is the number of pixels in the image, which would be the case if the image was all one color. We'll use the turtle module again here, but this example could be easily converted into any graphical output module. Let's take a look at the swap.tif image we created in the previous example:

1. First, we import the libraries we need, including the turtle graphics library:

```
from gdal import gdal_array
import turtle as t
```

2. Now, we create a histogram function that can take an array and sort the numbers into bins making up the histogram:

```
def histogram(a, bins=list(range(0, 256))):
  fa = a.flat
  n = gdal_array.numpy.searchsorted(gdal_array.numpy.sort(fa), bins)
  n = gdal_array.numpy.concatenate([n, [len(fa)]])
  hist = n[1:]-n[:-1]
  return hist
```

3. Finally, we have our turtle graphics function that takes a histogram and draws it:

```
def draw_histogram(hist, scale=True):
```

4. Draw the graph axes using the following code:

```
t.color("black")
axes = ((-355, -200), (355, -200), (-355, -200), (-355, 250))
t.up()
for p in axes:
  t.goto(p)
  t.down()
  t.up()
```

5. Then, we can label them:

```
t.goto(0, -250)
t.write("VALUE", font=("Arial, ", 12, "bold"))
t.up()
t.goto(-400, 280)
t.write("FREQUENCY", font=("Arial, ", 12, "bold"))
x = -355
y = -200
t.up()
```

6. Now, we'll add tick marks on the x-axis so that we can see the line values:

```
for i in range(1, 11):
    x = x+65
    t.goto(x, y)
    t.down()
    t.goto(x, y-10)
    t.up()
    t.goto(x, y-25)
    t.write("{}".format((i*25)), align="center")
```

7. We'll do the same for the y-axis:

```
x = -355
y = -200
t.up()
pixels = sum(hist[0])
if scale:
    max = 0
    for h in hist:
        hmax = h.max()
        if hmax > max:
            max = hmax
    pixels = max
label = int(pixels/10)
for i in range(1, 11):
    y = y+45
    t.goto(x, y)
    t.down()
    t.goto(x-10, y)
    t.up()
    t.goto(x-15, y-6)
    t.write("{}".format((i*label)), align="right")
```

8. We can begin plotting our histogram lines:

```
x_ratio = 709.0 / 256
y_ratio = 450.0 / pixels
colors = ["red", "green", "blue"]
for j in range(len(hist)):
  h = hist[j]
  x = -354
  y = -199
  t.up()
  t.goto(x, y)
  t.down()
  t.color(colors[j])
  for i in range(256):
    x = i * x_ratio
    y = h[i] * y_ratio
    x = x - (709/2)
    y = y + -199
    t.goto((x, y))
```

9. Finally, we can load our image and plot its histogram using the functions we defined previously:

```
im = "swap.tif"
histograms = []
arr = gdal_array.LoadFile(im)
for b in arr:
  histograms.append(histogram(b))
draw_histogram(histograms)
t.pen(shown=False)
t.done()
```

Here's what the histogram for `swap.tif` looks like after running the preceding code example:

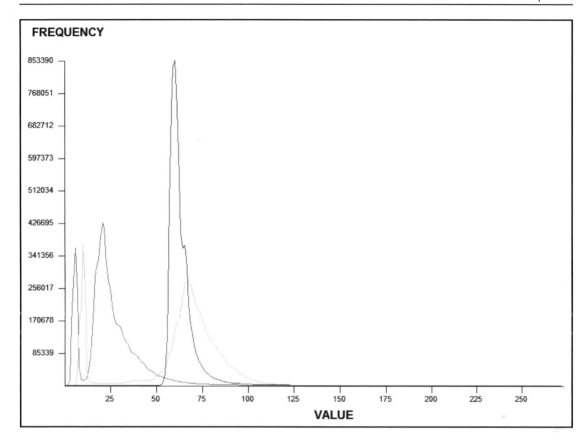

As you can see, all three bands are grouped closely toward the left-hand side of the graph and all have values less than **125** or so. As these values approach zero, the image becomes darker, which is not surprising.

Just for fun, let's run the script again and when we call the `draw_histogram()` function, we'll add the `scale=False` option to get a sense of the size of the image and provide an absolute scale. We'll change the following line:

```
draw_histogram(histograms)
```

This will be changed to the following:

```
draw_histogram(histograms, scale=False)
```

This change will produce the following histogram graph:

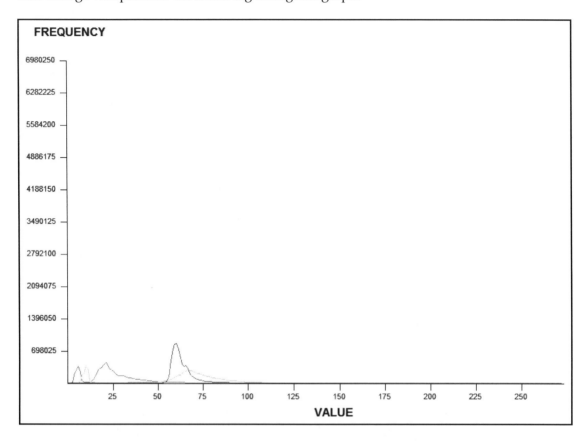

As you can see, it's harder to see the details of the value distribution. However, this absolute-scale approach is useful if you are comparing multiple histograms of different products that were produced from the same source image.

So, now that we understand the basics of looking at an image statistically using histograms, how do we make our image brighter? Let's check this out in the next section.

Performing a histogram stretch

A histogram stretch operation does exactly what its name says. It redistributes the pixel values across the whole scale. By doing so, we have more values at the higher-intensity level and the image becomes brighter. So, in this example, we'll reuse our histogram function, but we'll add another function called `stretch()` that takes an image array, creates the histogram, and then spreads out the range of values for each band. We'll run these functions on `swap.tif` and save the result in an image called `stretched.tif`:

```
import gdal_array
import operator
from functools import reduce

def histogram(a, bins=list(range(0, 256))):
 fa = a.flat
 n = gdal_array.numpy.searchsorted(gdal_array.numpy.sort(fa), bins)
 n = gdal_array.numpy.concatenate([n, [len(fa)]])
 hist = n[1:]-n[:-1]
 return hist

def stretch(a):
 """
 Performs a histogram stretch on a gdal_array array image.
 """
 hist = histogram(a)
 lut = []
 for b in range(0, len(hist), 256):
 # step size
 step = reduce(operator.add, hist[b:b+256]) / 255
 # create equalization look-up table
 n = 0
 for i in range(256):
 lut.append(n / step)
 n = n + hist[i+b]
 gdal_array.numpy.take(lut, a, out=a)
 return asrc = "swap.tif"
arr = gdal_array.LoadFile(src)
stretched = stretch(arr)
output = gdal_array.SaveArray(arr, "stretched.tif", format="GTiff",
prototype=src)
output = None
```

The `stretch` algorithm will produce the following image. Look how much brighter and visually appealing it is:

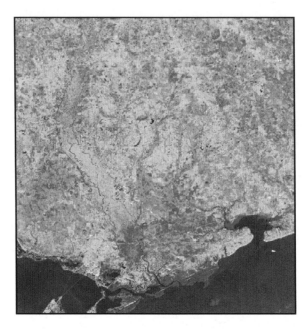

We can run our `turtle` graphics histogram script on `stretched.tif` by changing the file name in the `im` variable to `stretched.tif`:

```
im = "stretched.tif"
```

Running the preceding code will give us the following histogram:

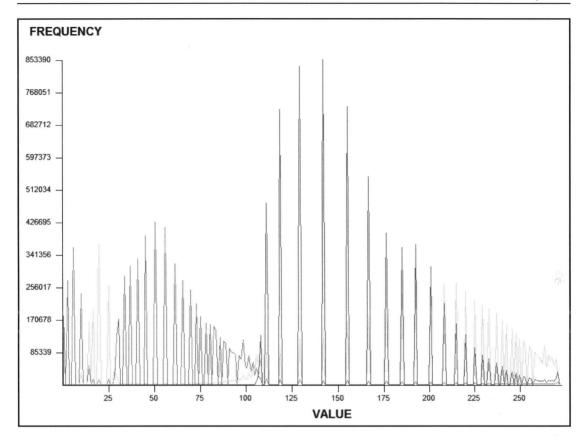

As you can see, all three bands are distributed evenly now. Their relative distribution to each other is the same, but, within the image, they are now spread across the spectrum.

Now that we can change images for better presentation, let's look at clipping them to examine a particular area of interest.

Clipping images

Very rarely is an analyst interested in an entire satellite scene, which can easily cover hundreds of square miles. Given the size of satellite data, we are highly motivated to reduce the size of an image to only our area of interest. The best way to accomplish this reduction is to clip an image to a boundary that defines our study area. We can use shapefiles (or other vector data) as our boundary definition and basically get rid of all the data outside that boundary.

The following image contains our `stretched.tif` image with a county boundary file layered on top, visualized in **Quantum GIS (QGIS)**:

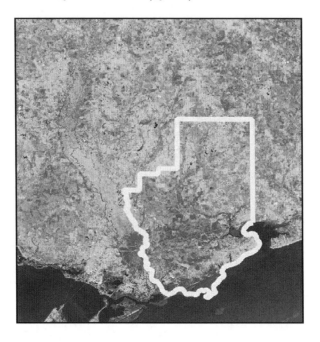

To clip the image, we need to follow these steps:

1. Load the image into an array using `gdal_array`.
2. Create a shapefile reader using PyShp.
3. Rasterize the shapefile into a georeferenced image (convert it from a vector into a raster).
4. Turn the shapefile image into a binary mask or filter to only grab the image pixels we want within the shapefile boundary.
5. Filter the satellite image through the mask.
6. Discard satellite image data outside the mask.
7. Save the clipped satellite image as `clip.tif`.

We installed PyShp in Chapter 4, *Geospatial Python Toolbox*, so you should already have it installed from PyPi. We will also add a couple of useful new utility functions in this script. The first is `world2pixel()`, which uses the GDAL GeoTransform object to do the world-coordinate to image-coordinate conversion for us.

It's still the same process we've used throughout this book, but it's integrated better with GDAL.

We also add the `imageToArray()` function, which converts a PIL image into a NumPy array. The county boundary shapefile is the `hancock.shp` boundary we've used in previous chapters, but you can also download it here if you need to: `http://git.io/vqsRH`.

We use PIL because it is the easiest way to rasterize our shapefile as a mask image to filter out the pixels beyond the shapefile boundary. Let's get started:

1. First, we'll load the libraries we need:

```
import operator
from osgeo import gdal, gdal_array, osr
import shapefile
```

2. Now, we'll load PIL. This may need to be installed slightly differently on different platforms, so we have to check for that difference:

```
try:
  import Image
  import ImageDraw
except:
  from PIL import Image, ImageDraw
```

3. Now, we will set up the variables for our input image, shapefile, and our output image:

```
# Raster image to clip
raster = "stretched.tif"
# Polygon shapefile used to clip
shp = "hancock"
# Name of clipped raster file(s)
output = "clip"
```

4. Next, create a function that simply converts an image into a numpy array so that we can convert the mask image we will create and use it in our NumPy-based clipping process:

```
def imageToArray(i):
  """
  Converts a Python Imaging Library array to a gdal_array image.
  """
  a = gdal_array.numpy.fromstring(i.tobytes(), 'b')
```

```
a.shape = i.im.size[1], i.im.size[0]
return a
```

5. Next, we need a function to convert geospatial coordinates into image pixels, which will allow us to use coordinates from our clipping shapefile to limit which image pixels are saved:

```
def world2Pixel(geoMatrix, x, y):
    """
    Uses a gdal geomatrix (gdal.GetGeoTransform()) to calculate
    the pixel location of a geospatial coordinate
    """
    ulX = geoMatrix[0]
    ulY = geoMatrix[3]
    xDist = geoMatrix[1]
    yDist = geoMatrix[5]
    rtnX = geoMatrix[2]
    rtnY = geoMatrix[4]
    pixel = int((x - ulX) / xDist)
    line = int((ulY - y) / abs(yDist))
    return (pixel, line)
```

6. Now, we can load our source image into a numpy array:

```
# Load the source data as a gdal_array array
srcArray = gdal_array.LoadFile(raster)
```

7. We'll also load the source image as a gdal image because gdal_array does not give us the geotransform information we need to convert coordinates into pixels:

```
# Also load as a gdal image to get geotransform (world file) info
srcImage = gdal.Open(raster)
geoTrans = srcImage.GetGeoTransform()
```

8. Now, we'll use the Python shapefile library to open our shapefile:

```
# Use pyshp to open the shapefile
r = shapefile.Reader("{}.shp".format(shp))
```

9. Next, we'll convert the shapefile bounding box coordinates into image coordinates based on our source image:

```
# Convert the layer extent to image pixel coordinates
minX, minY, maxX, maxY = r.bbox
ulX, ulY = world2Pixel(geoTrans, minX, maxY)
lrX, lrY = world2Pixel(geoTrans, maxX, minY)
```

10. Then, we can calculate the size of our output image based on the extents of the shapefile and take just that part of the source image:

```
# Calculate the pixel size of the new image
pxWidth = int(lrX - ulX)
pxHeight = int(lrY - ulY)
clip = srcArray[:, ulY:lrY, ulX:lrX]
```

11. Next, we'll create new geomatrix data for the output image:

```
# Create a new geomatrix for the image
# to contain georeferencing data
geoTrans = list(geoTrans)
geoTrans[0] = minX
geoTrans[3] = maxY
```

12. Now, we can create a simple black-and-white mask image from the shapefile that will define the pixels we want to extract from the source image:

```
# Map points to pixels for drawing the county boundary
# on a blank 8-bit, black and white, mask image.
pixels = []
for p in r.shape(0).points:
 pixels.append(world2Pixel(geoTrans, p[0], p[1]))
rasterPoly = Image.new("L", (pxWidth, pxHeight), 1)
# Create a blank image in PIL to draw the polygon.
rasterize = ImageDraw.Draw(rasterPoly)
rasterize.polygon(pixels, 0)
```

13. Next, we convert the mask image into a numpy array:

```
# Convert the PIL image to a NumPy array
mask = imageToArray(rasterPoly)
```

14. Finally, we're ready to use the mask array to clip the source array in numpy and save it to a new geotiff image:

```
# Clip the image using the mask
clip = gdal_array.numpy.choose(mask, (clip, 0)).astype(
 gdal_array.numpy.uint8)
# Save ndvi as tiff
gdal_array.SaveArray(clip, "{}.tif".format(output),
 format="GTiff", prototype=raster)
```

This script produces the following clipped image:

The areas that remain outside the county boundary that appear in black are actually called `NoData` values, meaning there is no information at that location, and are ignored by most geospatial software. Because images are rectangular, the `NoData` values are common for data that does not completely fill an image.

You have now walked through an entire workflow that is used by geospatial analysts around the world every day to prepare multispectral satellite and aerial images for use in a GIS. We'll look at how we can actually analyze images as information in the next section.

Classifying images

Automated remote sensing (**ARS**) is rarely ever done in the visible spectrum. ARS processes images without any human input. The most commonly available wavelengths outside of the visible spectrum are infrared and near-infrared.

The following illustration is a thermal image (band 10) from a fairly recent Landsat 8 flyover of the US Gulf Coast from New Orleans, Louisiana to Mobile, Alabama. The major natural features in the image have been labeled so that you can orient yourself:

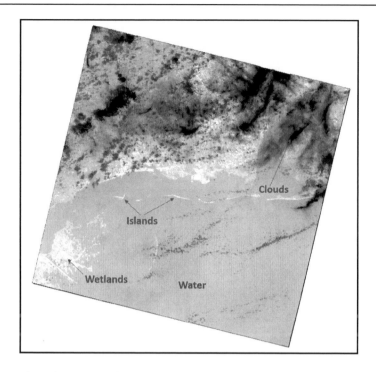

Because every pixel in that image has a reflectance value, it is information as opposed to just color. The type of reflectance can tell us definitively what a feature is, as opposed to us guessing by looking at it. Python can see those values and pick out features the same way we intuitively do by grouping related pixel values. We can colorize pixels based on their relation to each other to simplify the image and view-related features. This technique is called **classification**.

Classifying can range from fairly simple groupings, based only on some value distribution algorithm derived from the histogram, to complex methods involving training datasets and even computer learning and artificial intelligence. The simplest forms are called **unsupervised classifications**, in which no additional input is given other than the image itself. Methods involving some sort of training data to guide the computer are called **supervised classifications**. It should be noted that classification techniques are used across many fields, from medical doctors searching for cancerous cells in a patient's body scan, to casinos using facial-recognition software on security videos to automatically spot known **con-artists at blackjack tables**.

To introduce remote sensing classification, we'll just use the histogram to group pixels with similar colors and intensities and see what we get. First, you'll need to download the Landsat 8 scene from here: `http://git.io/vByJu`.

Instead of our `histogram()` function from the previous examples, we'll use the version included with NumPy that allows you to easily specify the number of bins and returns two arrays with the frequency, as well as the ranges of the bin values. We'll use the second array with the ranges as our class definitions for the image. The `lut` or look-up table is an arbitrary color palette that's used to assign colors to the 20 unsupervised classes. You can use any colors you want. Let's look at the following steps:

1. First, we import our libraries:

   ```
   import gdal
   from gdal import gdal_array, osr
   ```

2. Next, we set up some variables for our input and output images:

   ```
   # Input file name (thermal image)
   src = "thermal.tif"
   # Output file name
   tgt = "classified.jpg"
   ```

3. Load the image into a `numpy` array for processing:

   ```
   # Load the image into numpy using gdal
   srcArr = gdal_array.LoadFile(src)
   ```

4. Now, we're going to create a histogram of our image with 20 groups or `bins` that we'll use for classifying:

   ```
   # Split the histogram into 20 bins as our classes
   classes = gdal_array.numpy.histogram(srcArr, bins=20)[1]
   ```

5. Then, we'll create a look-up table that will define the color ranges for our classes so that we can visualize them:

   ```
   # Color look-up table (LUT) - must be len(classes)+1.
   # Specified as R, G, B tuples
   lut = [[255, 0, 0], [191, 48, 48], [166, 0, 0], [255, 64, 64],
   [255,
       115, 115], [255, 116, 0], [191, 113, 48], [255, 178, 115], [0,
       153, 153], [29, 115, 115], [0, 99, 99], [166, 75, 0], [0, 204,
       0], [51, 204, 204], [255, 150, 64], [92, 204, 204], [38, 153,
       38], [0, 133, 0], [57, 230, 57], [103, 230, 103], [184, 138,
   0]]
   ```

6. Now that our setup is complete, we can perform the classification:

```
# Starting value for classification
start = 1
# Set up the RGB color JPEG output image
rgb = gdal_array.numpy.zeros((3, srcArr.shape[0],
 srcArr.shape[1], ), gdal_array.numpy.float32)
# Process all classes and assign colors
for i in range(len(classes)):
  mask = gdal_array.numpy.logical_and(start <= srcArr, srcArr <=
  classes[i])
  for j in range(len(lut[i])):
    rgb[j] = gdal_array.numpy.choose(mask, (rgb[j], lut[i][j]))
  start = classes[i]+1
```

7. Finally, we can save our classified image:

```
# Save the image
output = gdal_array.SaveArray(rgb.astype(gdal_array.numpy.uint8),
tgt, format="JPEG")
output = None
```

The following image is our classification output, which we just saved as a JPEG:

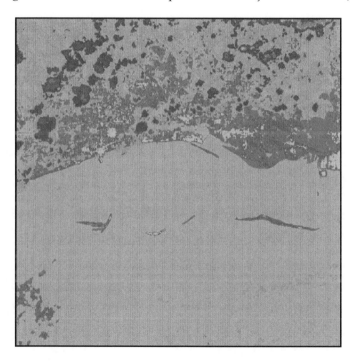

We didn't specify the prototype argument when saving this as an image, so it has no georeferencing information, though we could easily have done otherwise to save the output as a GeoTIFF.

This result isn't bad for a very simple unsupervised classification. The islands and coastal flats show up as different shades of green. The clouds were isolated as shades of orange and dark blues. We did have some confusion inland where the land features were colored the same as the Gulf of Mexico. We could further refine this process by defining the class ranges manually instead of just using the histogram.

Now that we have the ability to separate features in the image, we can try to extract features as vector data for inclusion in a GIS.

Extracting features from images

The ability to classify an image leads us to another remote sensing capability. Now that you've worked with shapefiles over the last few chapters, have you ever wondered where they come from? Vector GIS data such as shapefiles are typically extracted from remotely-sensed images such as the examples we've seen so far.

Extraction normally involves an analyst clicking around each object in an image and drawing the feature to save it as data. But with good remotely-sensed data and proper pre-processing, it is possible to automatically extract features from an image.

For this example, we'll take a subset of our Landsat 8 thermal image to isolate a group of barrier islands in the Gulf of Mexico. The islands appear white as the sand is hot and the cooler water appears black (you can download this image here: `http://git.io/vqarj`):

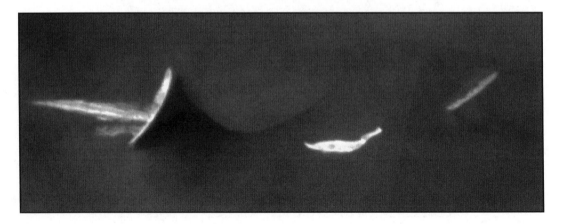

Our goal with this example is to automatically extract the three islands in the image as a shapefile. But before we can do that, we need to mask out any data we aren't interested in. For example, the water has a wide range of pixel values, as do the islands themselves. If we just want to extract the islands themselves, we need to push all the pixel values into just two bins to make the image black and white. This technique is called **thresholding**. The islands in the image have enough contrast with the water in the background that thresholding should isolate them nicely.

In the following script, we will read the image into an array and then histogram the image using only two bins. We will then use the colors black and white to color the two bins. This script is simply a modified version of our classification script with very limited output. Let's look at the following steps:

1. First, we import the one library we need:

    ```
    from gdal import gdal_array
    ```

2. Next, we define the variables for our input and output image:

    ```
    # Input file name (thermal image)
    src = "islands.tif"
    # Output file name
    tgt = "islands_classified.tiff"
    ```

3. Then, we can load the image:

    ```
    # Load the image into numpy using gdal
    srcArr = gdal_array.LoadFile(src)
    ```

4. Now, we can set up our simple classification scheme:

    ```
    # Split the histogram into 20 bins as our classes
    classes = gdal_array.numpy.histogram(srcArr, bins=2)[1]
    lut = [[255, 0, 0], [0, 0, 0], [255, 255, 255]]
    ```

5. Next, we classify the image:

    ```
    # Starting value for classification
    start = 1
    # Set up the output image
    rgb = gdal_array.numpy.zeros((3, srcArr.shape[0], srcArr.shape[1],
    ),
     gdal_array.numpy.float32)
    # Process all classes and assign colors
    for i in range(len(classes)):
      mask = gdal_array.numpy.logical_and(start <= srcArr, srcArr <=
      classes[i])
    ```

```
        for j in range(len(lut[i])):
            rgb[j] = gdal_array.numpy.choose(mask, (rgb[j], lut[i][j]))
            start = classes[i]+1
```

6. Finally, we save the image:

```
# Save the image
gdal_array.SaveArray(rgb.astype(gdal_array.numpy.uint8),
    tgt, format="GTIFF", prototype=src)
```

The output looks great, as shown in the following image:

The islands are clearly isolated, so our extraction script will be able to identify them as polygons and save them to a shapefile. The GDAL library has a method called `Polygonize()` that does exactly that. It groups all sets of isolated pixels in an image and saves them out as a feature dataset. One interesting technique we will use in this script is to use our input image as a mask.

The `Polygonize()` method allows you to specify a mask that will use the color black as a filter that will prevent the water from being extracted as a polygon, and we'll end up with just the islands. Another area to note in the script is that we copy the georeferencing information from our source image to our shapefile to geolocate it properly. Let's look at the following steps:

1. First, we import our libraries:

```
import gdal
from gdal import ogr, osr
```

2. Next, we set up our input and output image and shapefile variables:

```
# Thresholded input raster name
src = "islands_classified.tiff"
# Output shapefile name
```

```
tgt = "extract.shp"
# OGR layer name
tgtLayer = "extract"
```

3. Let's open our input image and get the first and only band:

```
# Open the input raster
srcDS = gdal.Open(src)
# Grab the first band
band = srcDS.GetRasterBand(1)
```

4. Then, we'll tell gdal to use that band as a mask:

```
# Force gdal to use the band as a mask
mask = band
```

5. Now, we're ready to set up our shapefile:

```
# Set up the output shapefile
driver = ogr.GetDriverByName("ESRI Shapefile")
shp = driver.CreateDataSource(tgt)
```

6. Then, we need to copy our spatial reference information from the source image to the shapefile, to locate it on the Earth:

```
# Copy the spatial reference
srs = osr.SpatialReference()
srs.ImportFromWkt(srcDS.GetProjectionRef())
layer = shp.CreateLayer(tgtLayer, srs=srs)
```

7. Now, we can set up our shapefile attributes:

```
# Set up the dbf file
fd = ogr.FieldDefn("DN", ogr.OFTInteger)
layer.CreateField(fd)
dst_field = 0
```

8. Finally, we can extract our polygons:

```
# Automatically extract features from an image!
extract = gdal.Polygonize(band, mask, layer, dst_field, [], None)
```

The output shapefile is simply called extract.shp. As you may remember from Chapter 4, *Geospatial Python Toolbox*, we created a quick pure Python script using PyShp and PNG Canvas to visualize shapefiles. We'll bring that script back here so that we can look at our shapefile, but we'll add something extra to it. The largest island has a small lagoon which shows up as a hole in the polygon. To properly render it, we have to deal with parts in a shapefile record.

The previous example using that script did not do that, so we'll add that piece as we loop through the shapefile features in the following steps:

1. First, we need to import the libraries we'll need:

   ```
   import shapefile
   import pngcanvas
   ```

2. Next, we get the spatial information from the shapefile that will allow us to map coordinates to pixels:

   ```
   r = shapefile.Reader("extract.shp")
   xdist = r.bbox[2] - r.bbox[0]
   ydist = r.bbox[3] - r.bbox[1]
   iwidth = 800
   iheight = 600
   xratio = iwidth/xdist
   yratio = iheight/ydist
   ```

3. Now, we'll create a list to hold our polygons:

   ```
   polygons = []
   ```

4. Then, we will loop through the shapefile and collect our polygons:

   ```
   for shape in r.shapes():
    for i in range(len(shape.parts)):
    pixels = []
    pt = None
    if i < len(shape.parts)-1:
      pt = shape.points[shape.parts[i]:shape.parts[i+1]]
    else:
      pt = shape.points[shape.parts[i]:]
   ```

5. Next, we map each point to an image pixel:

   ```
   for x, y in pt:
     px = int(iwidth - ((r.bbox[2] - x) * xratio))
     py = int((r.bbox[3] - y) * yratio)
     pixels.append([px, py])
   polygons.append(pixels)
   ```

6. Next, we draw the image using our polygon pixel information in PNGCanvas:

   ```
   c = pngcanvas.PNGCanvas(iwidth, iheight)
   for p in polygons:
    c.polyline(p)
   ```

7. Finally, we save the image:

```
with open("extract.png", "wb") as f:
    f.write(c.dump())
    f.close()
```

The following image shows our automatically extracted island features:

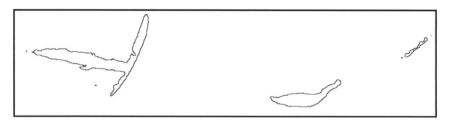

Commercial packages that do this kind of work can easily cost tens of thousands of dollars. While these packages are very robust, it is still fun and empowering to see how far you can get with simple Python scripts and a few open-source packages. In many cases, you can do everything you need to do.

The western-most island contains the polygon hole, as shown in the following image, and is zoomed in on that area:

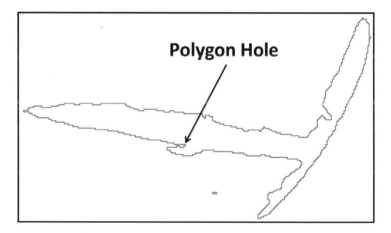

Polygon Hole

If you want to see what would happen if we didn't deal with the polygon holes, then just run the version of the script from Chapter 4, *Geospatial Python Toolbox*, on this same shapefile to compare the difference. The lagoon is not easy to see, but you will find it if you use the other script.

Automated feature extraction is a holy grail within geospatial analysis because of the cost and tedious effort required to manually extract features. The key to feature extraction is proper image classification. Automated feature extraction works well with water bodies, islands, roads, farm fields, buildings, and other features that tend to have high-contrast pixel values with their background.

You now have a good grasp of working with remote sensing data using GDAL, NumPy, and PIL. It's time to move on to our most complex example: change detection.

Understanding change detection

Change detection is the process of taking two geo-registered images of the exact same area from two different dates and automatically identifying differences. It is really just another form of image classification. Just like our previous classification examples, it can range from trivial techniques like those used here, to highly-sophisticated algorithms that provide amazingly precise and accurate results.

For this example, we'll use two images from a coastal area. These images show a populated area before and after a major hurricane, so there are significant differences, many of which are easy to visually spot, making these samples good for learning change detection. Our technique is to simply subtract the first image from the second to get a simple image difference using NumPy. This is a valid and often used technique.

The advantages are it is comprehensive and very reliable. The disadvantage of this overly simple algorithm is that it doesn't isolate the type of change. Many changes are insignificant for analysis, such as the waves on the ocean. In this example, we'll mask the water fairly effectively to avoid that distraction and only focus on the higher reflectance values toward the right-hand side of the difference image histogram.

You can download the baseline image from http://git.io/vqa6h. You can download the changed image from http://git.io/vqaic. Note these images are quite large – 24 MB and 64 MB, respectively!

The baseline image is panchromatic, while the changed image is in false color. Panchromatic images are created by sensors that capture all visible light and are typical of higher resolution sensors rather than multispectral sensors that capture bands containing restricted wavelengths.

Normally, you would use two identical band combinations, but these samples will work for our purposes. The visual markers we can use to evaluate change detection include a bridge in the southeast quadrant of the image that spans from the Peninsula to the edge of the image. This bridge is clearly visible in the before image and is reduced to pilings by the hurricane. Another marker is a boat in the northwest quadrant which appears in the after image as a white trail but is not in the before image.

A neutral marker is the water and the state highway, which runs through the town and connects to the bridge. This feature is easily visible concrete, which does not change significantly between the two images. The following is a screenshot of the baseline image:

To view these images up-close yourself, you should use QGIS or OpenEV (FWTools), as described in the *Quantum GIS and OpenEv* section in `Chapter 3`, *The Geospatial Technology Landscape*, to view them easily. The following image is the after image:

So, let's perform change detection:

1. First, we load our libraries:

```
import gdal
from gdal import gdal_array
import numpy as np
```

2. Now, we set up the variables for our input and output images:

```
# "Before" image
im1 = "before.tif"
# "After" image
im2 = "after.tif"
```

3. Next, we read both images into NumPy arrays with `gdal_array`:

```
# Load before and after into arrays
ar1 = gdal_array.LoadFile(im1).astype(np.int8)
ar2 = gdal_array.LoadFile(im2)[1].astype(np.int8)
```

4. Now, we subtract the before image from the after image (difference = after – before):

```
# Perform a simple array difference on the images
diff = ar2 - ar1
```

5. Then, we divide the image into five classes:

```
# Set up our classification scheme to try
# and isolate significant changes
classes = np.histogram(diff, bins=5)[1]
```

6. Next, we set our color table to use black to mask the lower classes. We do this to filter water and roads because they are darker in the image:

```
# The color black is repeated to mask insignificant changes
lut = [[0, 0, 0], [0, 0, 0], [0, 0, 0], [0, 0, 0], [0, 255, 0],
[255, 0, 0]]
```

7. Then, we assign colors to the classes:

```
# Starting value for classification
start = 1
# Set up the output image
rgb = np.zeros((3, diff.shape[0], diff.shape[1], ), np.int8)
# Process all classes and assign colors
for i in range(len(classes)):
 mask = np.logical_and(start <= diff, diff <= classes[i])
 for j in range(len(lut[i])):
 rgb[j] = np.choose(mask, (rgb[j], lut[i][j]))
 start = classes[i]+1
```

8. Finally, we save our image:

```
# Save the output image
output = gdal_array.SaveArray(rgb, "change.tif", format="GTiff",
prototype=im2)
output = None
```

Here's what our initial difference image looks like:

For the most part, the green classes represent areas where something was added. The red would be a darker value where something was probably removed. We can see that the boat trail is green in the northwest quadrant. We can also see a lot of changes in vegetation, as would be expected due to seasonal differences. The bridge is an anomaly because the exposed pilings are brighter than the darker surface of the original bridge, which makes them green instead of red.

Concrete is a major indicator in change detection because it is very bright in sunlight and is usually a sign of new development. Conversely, if a building is torn down and the concrete is removed, the difference is also easy to identify. So, the simple difference algorithm that we used here isn't perfect, but it could be greatly improved using thresholding, masking, better class definitions, and other techniques.

To really appreciate our change detection product, you can overlay it on the before or after image in QGIS and set the color black to transparent, as shown in the following image:

Potentially, you can combine this change detection analysis with the feature extraction example to extract changes as vector data that can be analyzed in a GIS efficiently.

Summary

In this chapter, we covered the foundations of remote sensing, including band swapping, histograms, image classification, feature extraction, and change detection. Like in the other chapters, we stayed as close to pure Python as possible, and where we compromised on this goal for processing speed, we limited the software libraries as much as possible to keep things simple. However, if you have the tools from this chapter installed, you really have a complete remote sensing package that is limited only by your desire to learn.

The techniques in this chapter are foundational to all remote sensing processes and will allow you to build more complex operations.

In the next chapter, we'll investigate elevation data. Elevation data doesn't fit squarely in GIS or remote sensing as it has elements of both types of processing.

Further reading

The authors of GDAL have a set of Python examples that cover a number of advanced topics that may be of interest to you. You can find them at `https://github.com/OSGeo/gdal/tree/master/gdal/swig/python/samples`.

Python and Elevation Data

7

Elevation data is one of the most fascinating types of geospatial data. It represents many different types of data sources and formats. It can display properties of both vector and raster data, resulting in unique data products. Elevation data can be used for terrain visualization, land cover classification, hydrology modeling, transportation routing, feature extraction, and many other purposes.

You can't perform all of these options with both raster and vector data, but since elevation data is three-dimensional, due to containing x, y, and z coordinates, you can often get more out of this data than any other type.

In this chapter, we will cover the following topics:

- Using ASCII Grid elevation data files for simple elevation processing
- Creating shaded relief images
- Creating elevation contours
- Gridding the LIDAR data
- Creating a 3D mesh

In this chapter, you will learn how to read and write elevation data in both raster and vector formats. We'll also create some derivative products.

Accessing ASCII Grid files

For most of this chapter, we'll use ASCII Grid files, or ASCIIGRID. These files are a type of raster data that's usually associated with elevation data. This grid format stores data as text in equal-sized square rows and columns with a simple header. Each cell in a row/column stores a single numeric value, which can represent some feature of terrain, such as elevation, slope, or flow direction. The simplicity makes it an easy-to-use and platform-independent raster format. This format is described in the *ASCII Grids* section of Chapter 2, *Learning Geospatial Data*.

Throughout this book, we've relied on GDAL, and to some extent, even PIL, to read and write geospatial raster data, including the `gdalnumeric` module, so that we can load raster data into NumPy arrays. ASCII Grid allows us to read and write rasters using only Python or even NumPy because it is simple plain text.

 As a reminder, some elevation datasets use image formats to store elevation data. Most image formats only support 8-bit values ranging from between 0 to 255; however, some formats, including TIFF, can store larger values.

Geospatial software can typically display these datasets; however, traditional image software and libraries usually don't. For simplicity, in this chapter, we'll mostly stick to the ASCII Grid format for data, which is both human and machine-readable, as well as widely supported.

Reading grids

NumPy has the ability to read the ASCII Grid format directly using its `loadtxt()` method, which is designed to read arrays from text files. The first six lines consist of the header, which is not a part of the array. The following lines are a sample of a grid header:

```
ncols 250
nrows 250
xllcorner 277750.0
yllcorner 6122250.0
cellsize 1.0
NODATA_value -9999
```

Let's look at what each line in the preceding code contains:

- Line 1 contains the number of columns in the grid, which is synonymous with the *x* axis.
- Line 2 represents the *y* axis as a number of rows.
- Line 3 represents the *x* coordinate of the lower-left corner, which is the minimum *x* value in meters.
- Line 4 is the corresponding minimum *y* value in the lower-left corner of the grid.
- Line 5 is the cell size or resolution of the raster. As the cells are square, only one size value is needed, as opposed to the separate *x* and *y* resolution values in most geospatial rasters.
- Line 6 is `NODATA_value`, which is a number that's assigned to any cell for which a value is not provided.

Geospatial software ignores these cells for calculations and often allows special display settings for it, such as making them black or transparent. The -9999 value is a common no data placeholder value that's used in the industry and is easy to detect in software but can be arbitrarily selected. Elevation with negative values (that is, bathymetry) may have valid data at -9999 meters, for instance, and may select 9999 or other values. As long as this value is defined in the header, most software will have no issues. In some examples, we'll use the number zero; however, zero can often also be a valid data value.

The `numpy.loadtxt()` method includes an argument called `skiprows`, which allows you to specify the number of lines in the file to be skipped before reading the array values.

 To try out this technique, you can download a sample grid file called `myGrid.asc` from `http://git.io/vYapU`.

So, for `myGrid.asc`, we would use the following code:

```
myArray  = numpy.loadtxt("myGrid.asc", skiprows=6)
```

This line results in the `myArray` variable containing a `numpy` array derived from the ASCIIGRID `myGrid.asc` file. The ASC filename extension is used by the ASCIIGRID format. This code works great, but there's one problem. NumPy allows us to skip the header but not keep it. We need to keep this so that we have a spatial reference for our data. We will also use it to save this grid or create a new one.

To solve this problem, we'll use Python's built-in `linecache` module to grab the header. We could open the file, loop through the lines, store each one in a variable, and then close the file. However, `linecache` reduces the solution to a single line. The following line reads the first line in the file to a variable called `line1`:

```
import linecache
line1 = linecache.getline("myGrid.asc", 1)
```

In the examples in this chapter, we'll use this technique to create a simple header processor that can parse these headers into Python variables in just a few lines. Now that we know how to read grids, let's learn how to write them.

Writing grids

Writing grids in NumPy is just as easy as reading them. We use the corresponding
numpy.savetxt() function to save a grid to a text file. The only catch is that we must
build and add the six lines of header information before we dump the array to the file. This
process is slightly different for different versions of NumPy. In either case, you build the
header as a string first. If you are using NumPy 1.7 or later, the savetext() method has an
optional argument called header that lets you specify a string as an argument. You can
quickly check your NumPy version from the command line using the following command:

```
python -c "import numpy;print(numpy.__version__)"
1.8.2
```

The backward-compatible method is to open a file, write the header, and then dump the
array. Here is a sample of the version 1.7 approach to save an array called myArray to an
ASCIIGRID file called myGrid.asc:

```
header = "ncols {}\n".format(myArray.shape[1])
header += "nrows {}\n".format(myArray.shape[0])
header += "xllcorner 277750.0\n"
header += "yllcorner 6122250.0\n"
header += "cellsize 1.0\n"
header += "NODATA_value -9999"
numpy.savetxt("myGrid.asc", myArray, header=header, fmt="%1.2f")
```

We make use of Python format strings, which allow you to put placeholders in a string to
format the Python objects to be inserted. The {} format variable turns the object you refer
to into a string. In this case, we are referencing the number of columns and rows in the
array.

In NumPy, an array has two properties:

- Size: It returns an integer for the number of values in the array.
- Shape: It returns a tuple with the number of rows and columns, respectively.

So, in the preceding example, we use the shape property tuple to add the row and column
counts to the header of our ASCII Grid. Notice that we also add a trailing newline character
for each line (\n). There is no reason to change the x and y values, cell size, or no data value
unless we altered them in the script.

The `savetxt()` method also has an `fmt` argument, which allows you to use Python format strings to specify how the array values are written. In this case, the `%1.2f` value specifies floats with at least one number and no more than two decimal places. The backward-compatible version for NumPy, before 1.6, builds the header string in the same way but creates the file handle first:

```
with open("myGrid.asc", "w") as f:
  f.write(header)
  numpy.savetxt(f, str(myArray), fmt="%1.2f")
```

As you'll see in the upcoming examples, this ability to produce valid geospatial data files using only NumPy is quite powerful. In the next couple of examples, we'll be using an ASCIIGRID **Digital Elevation Model (DEM)** of a mountainous area near Vancouver, British Columbia, in Canada.

 You can download this sample as a ZIP file at the following URL: `http://git.io/vYwUX`.

The following image is the raw DEM that was colorized using QGIS with a color ramp that makes the lower elevation values dark blue and higher elevation values bright red:

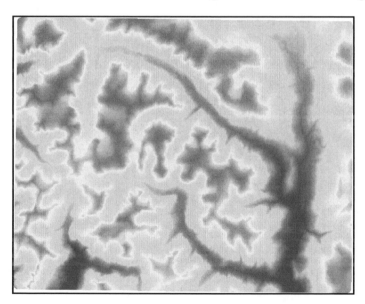

While we can conceptually understand the data in this way, it is not an intuitive way to visualize the data. Let's see if we can do better by creating a shaded relief.

Creating a shaded relief

Shaded relief maps color elevation in such a way that it looks as if the terrain is cast in a low angle light, which creates bright spots and shadows. This aesthetic styling creates an almost photographic illusion, which is easy to grasp so that we can understand the variation in the terrain. It is important to note that this style is truly an illusion as the light is often physically inaccurate in terms of the solar angle, and the elevation is usually exaggerated to increase contrast.

In this example, we'll use the ASCII DEM we referenced previously to create another grid that represents a shaded relief version of the terrain in NumPy. This terrain is quite dynamic, so we won't need to exaggerate the elevation; however, the script has a variable called z, which can be increased from 1.0 to scale the elevation up.

After we have defined all the variables, including the input and output filenames, we'll see the header parser based on the `linecache` module, which also uses a Python list comprehension to loop and parse the lines that are then split from a list into six variables. We also create a y cell size called `ycell`, which is just the inverse of the cell size by convention. If we don't do this, the resulting grid will be transposed.

Note that we define filenames for slope and aspect grids, which are two intermediate products that are combined to create the final product. These intermediate grids are output as well. They can also serve as inputs to other types of products.

This script uses a three-by-three windowing method to scan the image and smooth out the center value in these mini-grids to process the image efficiently. It does so within the memory constraints of your computer. However, because we are using NumPy, we can process the entire array at once via matrices, as opposed to using a lengthy series of nested loops. This technique is based on the excellent work of a developer named Michal Migurski, who implemented the clever NumPy version of Matthew Perry's C++ implementation, which served as the basis for the DEM tools in the GDAL suite.

After the slope and aspect have been calculated, they are used to output the shaded relief. The slope is the steepness of a hill or mountain, while the aspect is the direction the grid cell faces that is specified as a degree between 0 and 360. Finally, everything is saved to the disk from NumPy. In the `savetxt()` method, we specify a four-integer format string as the peak elevations are several thousand meters:

1. First, we'll import the `linecache` module to parse the header and the `numpy` module to do the processing:

```
from linecache import getline
import numpy as np
```

2. Next, we'll set up all of the variable names that will define how the shaded relief is processed:

```
# File name of ASCII digital elevation model
source = "dem.asc"

# File name of the slope grid
slopegrid = "slope.asc"

# File name of the aspect grid
aspectgrid = "aspect.asc"

# Output file name for shaded relief
shadegrid = "relief.asc"

# Shaded elevation parameters
# Sun direction
azimuth = 315.0

# Sun angle
altitude = 45.0

# Elevation exaggeration
z = 1.0

# Resolution
scale = 1.0

# No data value for output
NODATA = -9999

# Needed for numpy conversions
deg2rad = 3.141592653589793 / 180.0
rad2deg = 180.0 / 3.141592653589793
```

3. Now that our variables are set up, we can parse the header:

```
# Parse the header using a loop and
# the built-in linecache module
hdr = [getline(source, i) for i in range(1, 7)]
values = [float(h.split(" ")[-1].strip()) for h in hdr]
cols, rows, lx, ly, cell, nd = values
xres = cell
yres = cell * -1
```

4. Next, we can load the actual data using numpy by skipping the header portion:

```
# Load the dem into a numpy array
arr = np.loadtxt(source, skiprows=6)
```

5. We're going to loop through the data, row by row, column by column, to process it. Please note, however, that we're going to skip the outer edges that contain nodata values. We'll break the data into smaller grids of 3 x 3 pixels as we go because for each grid cell, we need to see the cells surrounding it:

```
# Exclude 2 pixels around the edges which are usually NODATA.
# Also set up structure for 3 x 3 windows to process the slope
# throughout the grid
window = []
for row in range(3):
 for col in range(3):
 window.append(arr[row:(row + arr.shape[0] - 2),
 col:(col + arr.shape[1] - 2)])

# Process each 3x3 window in both the x and y directions
x = ((z * window[0] + z * window[3] + z * window[3] + z *
 window[6]) -
 (z * window[2] + z * window[5] + z * window[5] + z *
 window[8])) / \
 (8.0 * xres * scale)
y = ((z * window[6] + z * window[7] + z * window[7] + z *
 window[8]) -
 (z * window[0] + z * window[1] + z * window[1] + z *
 window[2])) / \
 (8.0 * yres * scale)
```

6. For each 3 x 3 mini-window, we'll calculate slope, aspect, and then the shaded relief value:

```
# Calculate slope
slope = 90.0 - np.arctan(np.sqrt(x * x + y * y)) * rad2deg

# Calculate aspect
```

```
aspect = np.arctan2(x, y)

# Calculate the shaded relief
shaded = np.sin(altitude * deg2rad) * np.sin(slope * deg2rad) + \
  np.cos(altitude * deg2rad) * np.cos(slope * deg2rad) * \
  np.cos((azimuth - 90.0) * deg2rad - aspect)
```

7. Next, we need to scale each value between 0-255 so that it can be viewed as an image:

```
# Scale values from 0-1 to 0-255
shaded = shaded * 255
```

8. Now, we have to rebuild our header since we have ignored the outer edge of the nodata values and our dataset is smaller:

```
# Rebuild the new header
header = "ncols {}\n".format(shaded.shape[1])
header += "nrows {}\n".format(shaded.shape[0])
header += "xllcorner {}\n".format(lx + (cell * (cols -
  shaded.shape[1])))
header += "yllcorner {}\n".format(ly + (cell * (rows -
  shaded.shape[0])))
header += "cellsize {}\n".format(cell)
header += "NODATA_value {}\n".format(NODATA)
```

9. Next, we'll set any nodata values to the chosen nodata values we set in our variables at the beginning:

```
# Set no-data values
for pane in window:
  slope[pane == nd]  = NODATA
  aspect[pane == nd] = NODATA
  shaded[pane == nd] = NODATA
```

10. We're going to save the slope and aspect grids separately so that we can view them later and understand how the shaded relief is created:

```
# Open the output file, add the header, save the slope grid
with open(slopegrid, "wb") as f:
  f.write(bytes(header, "UTF-8")
  np.savetxt(f, slope, fmt="%4i")

# Open the output file, add the header, save the aspectgrid
with open(aspectgrid, "wb") as f:
  f.write(bytes(header, "UTF-8")
  np.savetxt(f, aspect, fmt="%4i")
```

```
# Open the output file, add the header, save the relief grid
with open(shadegrid, "wb") as f:
  f.write(bytes(header, 'UTF-8'))
  np.savetxt(f, shaded, fmt="%4i")
```

If we load the output shaded relief grid to QGIS and specify the styling to stretch the image to the minimum and maximum values, we will see the following image:

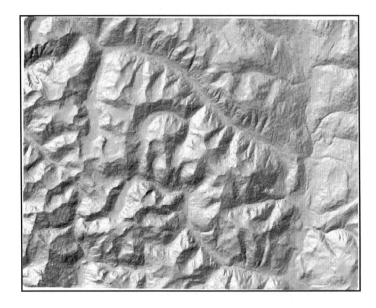

If QGIS asks you for a projection, the data is EPSG:3157. You can also open the image in the FWTools OpenEV application we discussed in the *Installing GDAL* section of Chapter 4, *Geospatial Python Toolbox*, which will automatically stretch the image for optimal viewing.

As you can see, the preceding image is much easier to comprehend than the pseudo-color representation that we examined originally. Next, let's look at the slope raster that's used to create the shaded relief:

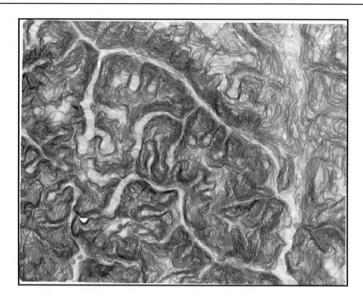

The slope shows the gradual decline in elevation from the high points to low points in all the directions of the dataset. The slope is an especially useful input for many types of hydrology models:

The aspect shows the maximum rate of a downslope change from one cell to its neighbors. If you compare the aspect image to the shaded relief image, you will see that the red and gray values of the aspect image correspond to shadows in the shaded relief. So, the slope is primarily responsible for turning the DEM into a terrain relief while the aspect is responsible for shading.

Now that we can display the data in a useful way, let's see if we can also create other data from it.

Creating elevation contours

A contour is an isoline along the same elevation in a dataset. Contours are usually stepped at intervals to create an intuitive way to represent elevation data, both visually and numerically, using a resource-efficient vector dataset. Now, let's look at another way to visualize the elevation better using contours.

The input is used to generate contours in our DEM and the output is a shapefile. The algorithm (Marching Squares: `https://en.wikipedia.org/wiki/Marching_squares`) that's used to generate contours is fairly complex and very difficult to implement using NumPy's linear algebra. In this case, our solution is to fall back on the GDAL library, which has a contouring method available through the Python API. In fact, the majority of this script is just setting up the OGR library code that is needed to output a shapefile. The actual contouring is a single method call named `gdal.ContourGenerate()`. Just before this call, there are comments that define the method's arguments. The most important ones are as follows:

- `contourInterval`: This is the distance in the dataset units between contours.
- `contourBase`: This is the starting elevation for the contouring.
- `fixedLevelCount`: This specifies a fixed number of contours as opposed to distance.
- `idField`: This is the name for a required shapefile `dbf` field, usually just called ID.
- `elevField`: This is the name for a required shapefile `dbf` field for the elevation value and is useful for labeling in maps.

You should have GDAL and OGR installed from the *Installing GDAL* section of `Chapter 4,` *Geospatial Python Toolbox*. We will be implementing the following steps:

1. First, we will define the input DEM filename.
2. Then, we will output the shapefile's name.
3. Next, we'll create the shapefile data source with OGR.
4. Then, we'll get the OGR layer.
5. Next, we'll open the DEM.
6. Finally, we'll generate contours on the OGR layer.

Let's look at a code representation of the preceding steps:

1. First, we load in the `gdal` and `ogr` libraries to handle the data:

```
import gdal
import ogr
```

2. Then we'll set up a variable for our filename:

```
# Elevation DEM
source = "dem.asc"
```

3. Next, we'll create the beginnings of our output shapefile using OGR:

```
# Output shapefile
target = "contour"
ogr_driver = ogr.GetDriverByName("ESRI Shapefile")
ogr_ds = ogr_driver.CreateDataSource(target + ".shp")
ogr_lyr = ogr_ds.CreateLayer(target,
# wkbLineString25D is the type code for geometry with a z
# elevation value.
geom_type=ogr.wkbLineString25D)
field_defn = ogr.FieldDefn("ID" ogr.OFTInteger)
ogr_lyr.CreateField(field_defn)
field_defn = ogr.FieldDefn("ELEV" ogr.OFTReal)
ogr_lyr.CreateField(field_defn)
```

4. Then, we'll create some contours:

```
# gdal.ContourGenerate() arguments
# Band srcBand,
# double contourInterval,
# double contourBase,
# double[] fixedLevelCount,
# int useNoData,
# double noDataValue,
```

```
# Layer dstLayer,
# int idField,
# int elevField
ds = gdal.Open(source)

# EPGS:3157
gdal.ContourGenerate(ds.GetRasterBand(1), 400, 10, [], 0, 0,
ogr_lyr, 0, 1))
ogr_ds = None
```

5. Now, let's draw the contour shapefile that we just created using pngcanvas, which we introduced in the *PNGCanvas* section of Chapter 4, *Geospatial Python Toolbox*:

```
import shapefile
import pngcanvas

# Open the contours
r = shapefile.Reader("contour.shp")

# Setup the world to pixels conversion
xdist = r.bbox[2] - r.bbox[0]
ydist = r.bbox[3] - r.bbox[1]
iwidth = 800
iheight = 600
xratio = iwidth/xdist
yratio = iheight/ydist
contours = []

# Loop through all shapes
for shape in r.shapes():
 # Loop through all parts
 for i in range(len(shape.parts)):
   pixels = []
   pt = None
   if i < len(shape.parts) - 1:
     pt = shape.points[shape.parts[i]:shape.parts[i+1]]
   else:
     pt = shape.points[shape.parts[i]:]
   for x, y in pt:
     px = int(iwidth - ((r.bbox[2] - x) * xratio))
     py = int((r.bbox[3] - y) * yratio)
     pixels.append([px, py])
     contours.append(pixels)

# Set up the output canvas
canvas = pngcanvas.PNGCanvas(iwidth, iheight)
```

```
# PNGCanvas accepts rgba byte arrays for colors
red = [0xff, 0, 0, 0xff]
canvas.color = red

# Loop through the polygons and draw them
for c in contours:
 canvas.polyline(c)

# Save the image
with open("contours.png", "wb") as f:
 f.write(canvas.dump())
```

We will end up with the following image:

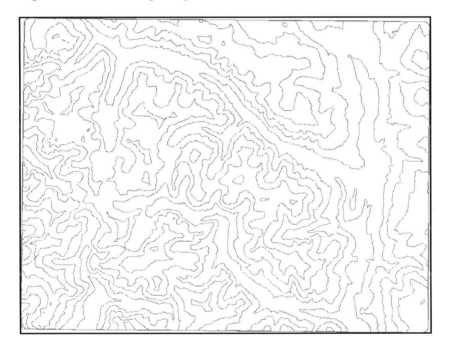

If we bring our shaded relief ASCIIGRID and the shapefile into a GIS, such as QGIS, we can create a simple topographic map, as follows. You can use the elevation (that is, `ELEV`) `dbf` field that you specified in the script to label the contour lines with the elevation:

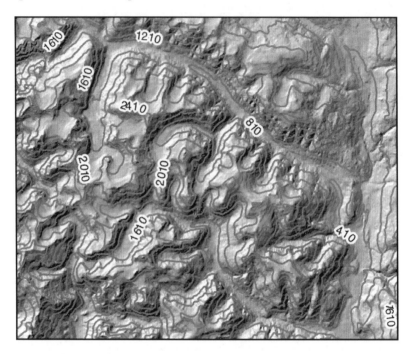

The techniques that were used in these NumPy grid examples provide the building blocks for all kinds of elevation products. Next, we'll work with one of the most complex elevation data types: LIDAR data.

Working with LIDAR data

LIDAR stands for **Light Detection and Ranging**. It is similar to radar-based images but uses finite laser beams that hit the ground hundreds of thousands of times per second to collect a huge amount of very fine (x,y,z) locations, as well as time and intensity. The intensity value is what really separates LIDAR from other data types. For example, the asphalt rooftop of a building may be of the same elevation as the top of a nearby tree, but the intensities will be different. Just like remote sensing, radiance values in a multispectral satellite image allow us to build classification libraries. The intensity values of LIDAR data allow us to classify and colorize LIDAR data.

The high volume and precision of LIDAR actually make it difficult to use. A LIDAR dataset is referred to as a point cloud because the shape of the dataset is usually irregular as the data is three-dimensional with outlying points. There's not many software packages that effectively visualize point clouds.

Furthermore, an irregular-shaped collection of finite points is just hard to interact with, even when we are using appropriate software.

For these reasons, one of the most common operations on LIDAR data is to project data and resample it to a regular grid. We'll do this using a small LIDAR dataset. This dataset is approximately 7 MB uncompressed and contains over 600,000 points. The data captures some easily identifiable features, such as buildings, trees, and cars in parking lots. You can download the zipped dataset from `http://git.io/vOERW`.

The file format is a very common binary format specific to LIDAR called **LAS**, which is short for laser. Unzip this file to your working directory. To read this format, we'll use a pure Python library called `laspy`. You can install Python version 3.7 using the following command:

```
pip install http://git.io/vOER9
```

With `laspy` installed, we are ready to create a grid from LIDAR.

Creating a grid from the LIDAR data

This script is fairly straightforward. We loop through the (x,y) point locations in the LIDAR data and project them to our grid with a cell size of one meter. Due to the precision of the LIDAR data, we'll end up with multiple points in a single cell. We average these points to create a common elevation value. Another issue that we have to deal with is data loss. Whenever you resample the data, you lose information.

In this case, we'll end up with NODATA holes in the middle of the raster. To deal with this issue, we fill these holes with average values from the surrounding cells, which is a form of interpolation. We only need two modules, both available on PyPI, as shown in the following code:

```
from laspy.file import File
import numpy as np

# Source LAS file
source = "lidar.las"

# Output ASCII DEM file
```

```
target = "lidar.asc"

# Grid cell size (data units)
cell = 1.0

# No data value for output DEM
NODATA = 0

# Open LIDAR LAS file
las = File(source, mode="r")

# xyz min and max
min = las.header.min
max = las.header.max

# Get the x axis distance in meters
xdist = max[0] - min[0]

# Get the y axis distance in meters
ydist = max[1] - min[1]

# Number of columns for our grid
cols = int(xdist) / cell

# Number of rows for our grid
rows = int(ydist) / cell
cols += 1
rows += 1

# Track how many elevation
# values we aggregate
count = np.zeros((rows, cols)).astype(np.float32)

# Aggregate elevation values
zsum = np.zeros((rows, cols)).astype(np.float32)

# Y resolution is negative
ycell = -1 * cell

# Project x, y values to grid
projx = (las.x - min[0]) / cell
projy = (las.y - min[1]) / ycell

# Cast to integers and clip for use as index
ix = projx.astype(np.int32)
iy = projy.astype(np.int32)

# Loop through x, y, z arrays, add to grid shape,
```

```
# and aggregate values for averaging
for x, y, z in np.nditer([ix, iy, las.z]):
 count[y, x] += 1
 zsum[y, x] += z

# Change 0 values to 1 to avoid numpy warnings,
# and NaN values in array
nonzero = np.where(count > 0, count, 1)

# Average our z values
zavg = zsum / nonzero

# Interpolate 0 values in array to avoid any
# holes in the grid
mean = np.ones((rows, cols)) * np.mean(zavg)
left = np.roll(zavg, -1, 1)
lavg = np.where(left > 0, left, mean)
right = np.roll(zavg, 1, 1)
ravg = np.where(right > 0, right, mean)
interpolate = (lavg + ravg) / 2
fill = np.where(zavg > 0, zavg, interpolate)

# Create our ASCII DEM header
header = "ncols {}\n".format(fill.shape[1])
header += "nrows {}\n".format(fill.shape[0])
header += "xllcorner {}\n".format(min[0])
header += "yllcorner {}\n".format(min[1])
header += "cellsize {}\n".format(cell)
header += "NODATA_value {}\n".format(NODATA)

# Open the output file, add the header, save the array
with open(target, "wb") as f:
 f.write(bytes(header, 'UTF-8'))
 # The fmt string ensures we output floats
 # that have at least one number but only
 # two decimal places
 np.savetxt(f, fill, fmt="%1.2f")
```

The result of our script is an ASCIIGRID, which looks like the following image when viewed in OpenEV. Higher elevations are lighter while lower elevations are darker. Even in this form, you can see buildings, trees, and cars:

If we assigned a heat map color ramp, the colors give you a sharper sense of the elevation differences:

So, what happens if we run this output DEM through our shaded relief script from earlier? There's a big difference between straight-sided buildings and sloping mountains. If you change the input and output names in the shaded relief script to process the LIDAR DEM, we get the following slope result:

The gently rolling slope of the mountainous terrain is reduced to outlines of major features in the image. In the aspect image, the changes are so sharp and over such short distances that the output image is very chaotic to view, as shown in the following screenshot:

Despite the difference between these images and the coarser but somewhat smoother mountain versions, we still get a very nice shaded relief, which visually resembles a black and white photograph:

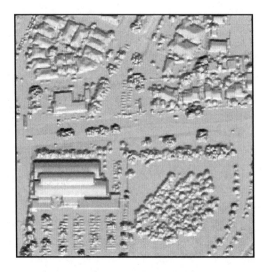

Now that we know how to process LIDAR data, let's learn how to visualize it using Python.

Using PIL to visualize LIDAR data

The previous DEM images in this chapter were visualized using QGIS and OpenEV. We can also create output images in Python by introducing some new functions of the **Python Imaging Library** (**PIL**) that we didn't use in the previous chapters.

In this example, we'll use the PIL.ImageOps module, which has functions for histogram equalization and automatic contrast enhancement. We'll use PIL's fromarray() method to import the data from numpy. Let's see how close we can get to the output of the desktop GIS programs that were pictured in this chapter with the help of the following code:

```python
import numpy as np

try:
  import Image
  import ImageOps
except ImportError:
  from PIL import Image, ImageOps

# Source gridded LIDAR DEM file
```

```
source = "lidar.asc"

# Output image file
target = "lidar.bmp"

# Load the ASCII DEM into a numpy array
arr = np.loadtxt(source, skiprows=6)

# Convert array to numpy image
im = Image.fromarray(arr).convert("RGB")

# Enhance the image:
# equalize and increase contrast
im = ImageOps.equalize(im)
im = ImageOps.autocontrast(im)

# Save the image
im.save(target)
```

As you can see, in the following image, the enhanced shaded relief has sharper relief than the previous version:

Now, let's colorize our shaded relief. We'll use the built-in Python `colorsys` module for color space conversion. Normally, we specify colors as RGB values. However, to create a color ramp for a heat map scheme, we'll use **HSV** (short for **Hue, Saturation, and Value**) values to generate our colors.

The advantage of HSV is that you can tweak the *H* value to be a degree between 0 and 360 on a color wheel. Using a single value for hue allows you to use a linear ramping equation, which is much easier than trying to deal with combinations of three separate RGB values. The following image, which was taken from the online magazine *Qt Quarterly*, illustrates the HSV color model:

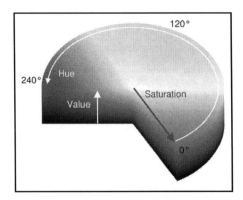

The `colorsys` module lets you switch back and forth between the HSV and RGB values. The module returns percentages for RGB values, which must then be mapped to the 0-255 scale for each color.

In the following code, we'll convert the ASCII DEM into a PIL image, build our color palette, apply the color palette to the grayscale image, and save the image:

```python
import numpy as np

try:
  import Image
  import ImageOps
except:
  from PIL import Image, ImageOps
import colorsys

# Source LIDAR DEM file
source = "lidar.asc"

# Output image file
target = "lidar.bmp"

# Load the ASCII DEM into a numpy array
arr = np.loadtxt(source, skiprows=6)

# Convert the numpy array to a PIL image.
# Use black and white mode so we can stack
```

```
# three bands for the color image.
im = Image.fromarray(arr).convert('L')

# Enhance the image
im = ImageOps.equalize(im)
im = ImageOps.autocontrast(im)

# Begin building our color ramp
palette = []

# Hue, Saturation, Value
# color space starting with yellow.
h = .67
s = 1
v = 1

# We'll step through colors from:
# blue-green-yellow-orange-red.
# Blue=low elevation, Red=high-elevation
step = h / 256.0

# Build the palette
for i in range(256):
 rp, gp, bp = colorsys.hsv_to_rgb(h, s, v)
 r = int(rp * 255)
 g = int(gp * 255)
 b = int(bp * 255)
 palette.extend([r, g, b])
 h -= step

# Apply the palette to the image
im.putpalette(palette)

# Save the image
im.save(target)
```

The preceding code produces the following image, with higher elevations in warmer colors and lower elevations in cooler colors:

In this image, we actually get more variation than the default QGIS version. We could potentially improve this image with a smoothing algorithm that would blend the colors where they meet and soften the image visually. As you can see, we have the full range of our color ramp expressed from cool to warm colors, as the elevation change increases.

Creating a triangulated irregular network

The following example is our most sophisticated example yet. A **triangulated irregular network (TIN)** is a vector representation of a point dataset in a vector surface of points connected as triangles. An algorithm determines which points are absolutely necessary to accurately represent the terrain as opposed to a raster, which stores a fixed number of cells over a given area and may repeat elevation values in adjacent cells that could be more efficiently stored as a polygon.

A TIN can also be resampled more efficiently on the fly than a raster, which requires less computer memory and processing power when using TIN in a GIS. The most common type of TIN is based on **Delaunay triangulation**, which includes all the points without redundant triangles.

The Delaunay triangulation is very complex. We'll use a pure Python library written by Bill Simons as part of Steve Fortune's Delaunay triangulation algorithm called `voronoi.py` to calculate the triangles in our LIDAR data. You can download the script to your working directory or `site-packages` directory from `http://git.io/vOEuJ`.

This script reads the LAS file, generates the triangles, loops through them, and writes out a shapefile. For this example, we'll use a clipped version of our LIDAR data to reduce the area to process. If we run our entire dataset of 600,000+ points, the script will run for hours and generate over half a million triangles. You can download the clipped LIDAR dataset as a ZIP file from the following URL: `http://git.io/vOE62`.

We have several status messages that print while the script runs because of the time-intensive nature of the following example, which can take several minutes to complete. We'll be storing the triangles as **PolygonZ types**, which allow the vertices to have a z elevation value. Unzip the LAS file and run the following code to generate a shapefile called `mesh.shp`:

1. First, we import our libraries:

```
import pickle
import os
import time
import math
import numpy as np
import shapefile
from laspy.file import File
# voronoi.py for Python 3: pip install http://git.io/vOEuJ
import voronoi
```

2. Next, we define the location and name of our LIDAR file, our target output file, and our pickle file:

```
# Source LAS file
source = "clippedLAS.las"

# Output shapefile
target = "mesh"

# Triangles pickle archive
archive = "triangles.p"
```

3. Now, we'll create a point class that's needed by the `voronoi` module:

```
class Point:
 """Point class required by the voronoi module"""
 def __init__(self, x, y):
   self.px = x
   self.py = y

 def x(self):
  return self.px

 def y(self):
  return self.py
```

4. Next, we'll create a triangle array to keep track of the triangles that have been created for the mesh:

```
# The triangle array holds tuples
# 3 point indices used to retrieve the points.
# Load it from a pickle
# file or use the voronoi module
# to create the triangles.
triangles = None
```

5. Next, we need to open our LIDAR file and pull the points:

```
# Open LIDAR LAS file
las = File(source, mode="r")
else:

# Open LIDAR LAS file
las = File(source, mode="r")
points = []
print("Assembling points...")

# Pull points from LAS file
for x, y in np.nditer((las.x, las.y)):
points.append(Point(x, y))
print("Composing triangles...")
```

6. Now, we can perform a Delaunay calculation on the points to build the triangles:

```
# Delaunay Triangulation
triangles = voronoi.computeDelaunayTriangulation(points)
```

7. We'll dump the triangles to the pickle archive to save time if we run this exact script again:

```
# Save the triangles to save time if we write more than
# one shapefile.
f = open(archive, "wb")
pickle.dump(triangles, f, protocol=2)
f.close()
```

8. Next, we can create a shapefile `Writer` object to begin creating our output shapefile by setting up the necessary fields:

```
print("Creating shapefile...")
# PolygonZ shapefile (x, y, z, m)
w = shapefile.Writer(target, shapefile.POLYGONZ)
w.field("X1", "C", "40")
w.field("X2", "C", "40")
w.field("X3", "C", "40")
w.field("Y1", "C", "40")
w.field("Y2", "C", "40")
w.field("Y3", "C", "40")
w.field("Z1", "C", "40")
w.field("Z2", "C", "40")
w.field("Z3", "C", "40")
tris = len(triangles)
```

9. Then, we loop through the triangles and create the mesh:

```
# Loop through shapes and
# track progress every 10 percent
last_percent = 0
for i in range(tris):
    t = triangles[i]
    percent = int(((i/(tris*1.0))*100.0)
    if percent % 10.0 == 0 and percent > last_percent:
        last_percent = percent
        print("{} % done - Shape {}/{} at {}".format(percent,
        i, tris, time.asctime()))
part = []
x1 = las.x[t[0]]
y1 = las.y[t[0]]
z1 = las.z[t[0]]
x2 = las.x[t[1]]
```

```
y2 = las.y[t[1]]
z2 = las.z[t[1]]
x3 = las.x[t[2]]
y3 = las.y[t[2]]
z3 = las.z[t[2]]
```

10. Next, we can eliminate any extremely long line segments, which are miscalculations by the library:

```
# Check segments for large triangles
# along the convex hull which is a common
# artifact in Delaunay triangulation
max = 3
if math.sqrt((x2-x1)**2+(y2-y1)**2) > max:
continue
if math.sqrt((x3-x2)**2+(y3-y2)**2) > max:
continue
if math.sqrt((x3-x1)**2+(y3-y1)**2) > max:
continue
part.append([x1, y1, z1, 0])
part.append([x2, y2, z2, 0])
part.append([x3, y3, z3, 0])
w.poly(parts=[part])
w.record(x1, x2, x3, y1, y2, y3, z1, z2, z3)
print("Saving shapefile...")
```

11. Finally, we can save the output shapefile:

```
w.close()
print("Done.")
```

The following image shows a zoomed-in version of the TIN over the colorized LIDAR data:

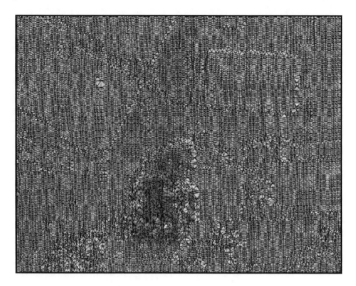

The mesh provides an efficient, continuous surface from point clouds, which can be easier to deal with than the point clouds themselves.

Summary

Elevation data can often provide a complete dataset for analysis and derivative products without any other data. In this chapter, you learned how to read and write ASCII Grids using only NumPy. You also learned how to create shaded reliefs, slope grids, and aspect grids. We created elevation contours using a little-known feature called contour of the GDAL library that's available for Python.

Next, we transformed LIDAR data into an easy-to-manipulate ASCII Grid. We experimented with different ways to visualize the LIDAR data with the PIL. Finally, we created a 3D surface or TIN by turning a LIDAR point cloud into a 3D shapefile of polygons. These are the tools of terrain analysis that are used for transportation planning, construction planning, hydrological drainage modeling, geologic exploration, and more.

In the next chapter, we'll combine the building blocks from the previous three chapters to perform some advanced modeling and actually create some information products.

Further reading

You can find some additional tutorials on Python and elevation data at the following link: `https://www.earthdatascience.org/tutorials/python/elevation/`.

Section 3: Practical Geospatial Processing Techniques

3

This section is of an advanced level and it will require all of the skills you learned previously. It starts off with learning how to create geospatial models to answer specific questions. Further, it will show you a few techniques for building geospatial models and how it will help to predict the future using visualization concepts. We'll move on to accessing and processing real-time data. At the end of this section, we'll combine all that we learned in the previous sections and implement a system to create an outdoor running or hiking report based on GPS data and geotagged photos.

This section includes the following chapters:

- Chapter 8, *Advanced Geospatial Python Modeling*
- Chapter 9, *Real-Time Data*
- Chapter 10, *Putting It All Together*

Advanced Geospatial Python Modeling

8

In this chapter, we'll build on the data processing concepts that we've learned in order to create some full-scale information products. The previously introduced data processing methods rarely provide answers to questions by themselves. You combine these data processing methods to build a geospatial model from multiple processed datasets. A geospatial model is a simplified representation of some aspect of the real world, which helps us answer one or more questions about a project or problem. In this chapter, we will introduce some important geospatial algorithms that are commonly used in agriculture, emergency management, logistics, and other industries.

The products that we will create are as follows:

- A crop health map
- A flood inundation model
- A colorized hillshade
- A terrain routing map
- A street routing map
- A shapefile with links to geolocated photos

While these products are task-specific, the algorithms that are used to create them are widely applied in geospatial analysis. We will be covering the following topics in this chapter:

- Creating a normalized difference vegetative index (NVDI)
- Creating a flood inundation model
- Creating a color hillshade
- Performing least cost path analysis
- Converting the route to a shapefile

- Routing along streets
- Geolocating photos
- Calculating satellite image cloud cover

The examples in this chapter are longer and more involved than in the previous chapters. For that reason, there are far more code comments to make the programs easier to follow. We will also use more functions in these examples. In previous chapters, functions were mostly avoided for clarity, but these examples are sufficiently complex that certain functions make the code easier to read. These examples are actual processes that you would use on the job as a geospatial analyst.

Technical requirements

For this chapter, the following requirements need to be satisfied:

- **Version**: Python 3.6 or higher
- **RAM**: Minimum 6 GB (Windows), 8 GB (macOS); recommended 8 GB
- **Storage**: Minimum 7,200 RPM SATA with 20 GB of available space, recommended SSD with 40 GB of available space.
- **Processor**: Minimum Intel Core i3 2.5 GHz, recommended Intel Core i5.

Creating a normalized difference vegetative index

Our first example will be an **normalized difference vegetative index** (**NVDI**). NDVIs are used to show the relative health of plants in an area of interest. An NDVI algorithm uses satellite or aerial imagery to show relative health by highlighting the chlorophyll density in plants. NDVIs use only the red and near-infrared bands. The formula of NDVI is as follows:

```
NDVI = (Infrared - Red) / (Infrared + Red)
```

The goal of this analysis is to produce, to begin with, a multispectral image containing infrared and red bands, and end up with a pseudo color image using seven classes, which color the healthier plants darker green, less-healthy plants lighter green, and bare soil brown.

Because the health index is relative, it is important to localize the area of interest. You could perform a relative index for the entire globe but vast areas, such as the Sahara desert on the low-vegetation extreme and densely forested areas, such as the Amazon jungle, skew the results for vegetation in the middle range. However, that being said, climate scientists routinely create global NDVIs to study worldwide trends. The more common application, though, is for managed areas, such as a forest or a farm field, as in this example.

We will begin with an analysis of a single farm field in the Mississippi Delta. To do so, we'll start with a multispectral image of a fairly large area and use a shapefile in order to isolate a single field. The image in the following screenshot is our broad area, with the field of interest highlighted in yellow:

You can download this image and the shapefile for the farm field as a ZIP file from `http://git.io/v3fS9`.

For this example, we'll use GDAL, OGR, `gdal_array`/`numpy`, and the **Python Imaging Library** (**PIL**) to clip and process the data. In the other examples in this chapter, we'll just use simple ASCII Grids and NumPy. As we'll be using ASCII elevation grids, GDAL isn't required. In all examples, the scripts use the following convention:

- Import libraries.
- Define functions.
- Define global variables, such as filenames.
- Execute the analysis.
- Save the output.

Our approach to the crop health example is split into two scripts. The first script creates the index image, which is a grayscale image. The second script classifies the index and outputs a colored image. In this first script, we'll execute the following steps to create the index image:

1. Read the infrared band.
2. Read the field boundary shapefile.
3. Rasterize the shapefile to an image.
4. Convert the shapefile image to a NumPy array.
5. Use the NumPy array to clip the red band to the field.
6. Do the same for the infrared band.
7. Use the band arrays to execute the NDVI algorithm in NumPy.
8. Save the resulting indexing algorithm to a GeoTIFF file using `gdal_array`.

We will discuss this script in sections to make it easier to follow. The code comments will also tell you what is going on at each step of the way.

Setting up the framework

Setting up the framework will help us to import the modules that we need and set up the functions that we'll use for steps 1 to 5 of the preceding instructions. The `imageToArray()` function converts a PIL image to a NumPy array and is dependent on the `gdal_array` and PIL modules. The `world2Pixel()` function converts geospatial coordinates to the pixel coordinates of our target image. This function uses the georeferencing information that is presented by the `gdal` module. The `copy_geo()` function copies the georeferencing information from our source image to our target array but accounts for the offset that is created when we clip the image. These functions are fairly generic and can serve a role in a variety of different remote sensing processes beyond this example:

1. First, we import our libraries:

```
import gdal
from osgeo import gdal
from osgeo import gdal_array
from osgeo import ogr
try:
  import Image
  import ImageDraw
except ImportError:
  from PIL import Image, ImageDraw
```

2. Then, we need a function to convert an image to a `numpy` array:

```
def imageToArray(i):
    """
    Converts a Python Imaging Library
    array to a gdal_array image.
    """
    a = gdal_array.numpy.fromstring(i.tobytes(), 'b')
    a.shape = i.im.size[1], i.im.size[0]
    return a
```

3. Now, we'll set up a function to convert the coordinates to image pixels:

```
def world2Pixel(geoMatrix, x, y):
    """
    Uses a gdal geomatrix (gdal.GetGeoTransform())
    to calculate the pixel location of a
    geospatial coordinate
    """
    ulX = geoMatrix[0]
    ulY = geoMatrix[3]
    xDist = geoMatrix[1]
    yDist = geoMatrix[5]
    rtnX = geoMatrix[2]
    rtnY = geoMatrix[4]
    pixel = int((x - ulX) / xDist)
    line = int((ulY - y) / abs(yDist))
    return (pixel, line)
```

4. Finally, we'll create a function to copy geographic metadata from an image:

```
def copy_geo(array, prototype=None, xoffset=0, yoffset=0):
    """Copy geotransfrom from prototype dataset to array but account
    for x, y offset of clipped array."""
    ds = gdal_array.OpenArray(array)
    prototype = gdal.Open(prototype)
    gdal_array.CopyDatasetInfo(prototype, ds,
    xoff=xoffset, yoff=yoffset)
    return ds
```

The next step is to load the data, which we'll be checking in the next section.

Loading the data

In this section, we load the source image of a farm field using `gdal_array`, which takes it straight into a NumPy array. We also define the name of our output image, which will be `ndvi.tif`. One interesting piece of this section is that we load the source image a second time using the `gdal` module, as opposed to `gdal_array`.

This second call is to capture the georeferencing data for the image that is available through `gdal`, and not `gdal_array`. Fortunately, `gdal` only loads raster data on demand, so this approach avoids loading the complete dataset into the memory twice. Once we have the data as a multidimensional NumPy array, we split out the red and infrared bands, as they will both be used in the NDVI equation:

```
# Multispectral image used
# to create the NDVI. Must
# have red and infrared
# bands
source = "farm.tif"

# Output geotiff file name
target = "ndvi.tif"

# Load the source data as a gdal_array array
srcArray = gdal_array.LoadFile(source)

# Also load as a gdal image to
# get geotransform info
srcImage = gdal.Open(source)
geoTrans = srcImage.GetGeoTransform()

# Red and infrared (or near infrared) bands
r = srcArray[1]
ir = srcArray[2]
```

Now that we have our data loaded, we can turn our shapefile into a raster.

Rasterizing the shapefile

This section begins the process of clipping. However, the first step is to rasterize the shapefile that outlines the boundary of the specific area that we are going to analyze. That area is within the larger `field.tif` satellite image. In other words, we convert it from vector data to raster data. But we also want to fill in the polygon when we convert it so that it can be used as an image mask. The pixels in the mask will be correlated to the pixels in the red and infrared arrays.

Any pixels outside the mask will be turned to NODATA pixels so they are not processed as part of the NDVI. To make this correlation, we'll need the solid polygon to be a NumPy array, just like the raster bands. This approach will make sure our NDVI calculation will be limited to the farm field.

The easiest way to convert the shapefile polygon into a filled polygon as a NumPy array is to plot it as a polygon in a PIL image, fill that polygon in, and then convert it to a NumPy array using existing methods, in both PIL and NumPy, which allow that conversion.

In this example, we use the ogr module to read the shapefile, because we already have GDAL available. But, we could have also used PyShp to read the shapefile just as easily. If our farm field image was available as an ASCII Grid, we could have avoided using the gdal, gdal_array, and ogr modules altogether:

1. First, we open our shapefile and select the one and only layer:

```
# Clip a field out of the bands using a
# field boundary shapefile

# Create an OGR layer from a Field boundary shapefile
field = ogr.Open("field.shp")
# Must define a "layer" to keep OGR happy
lyr = field.GetLayer("field")
```

2. There's only one polygon, so we'll grab that feature:

```
# Only one polygon in this shapefile
poly = lyr.GetNextFeature()
```

3. Now we'll convert the layer extent to image pixel coordinates:

```
# Convert the layer extent to image pixel coordinates
minX, maxX, minY, maxY = lyr.GetExtent()
ulX, ulY = world2Pixel(geoTrans, minX, maxY)
lrX, lrY = world2Pixel(geoTrans, maxX, minY)
```

4. Then, we calculate the pixel size of the new image:

```
# Calculate the pixel size of the new image
pxWidth = int(lrX - ulX)
pxHeight = int(lrY - ulY)
```

5. Next, we create a new blank image at the correct size:

```
# Create a blank image of the correct size
# that will serve as our mask
clipped = gdal_array.numpy.zeros((3, pxHeight, pxWidth),
 gdal_array.numpy.uint8)
```

6. Now, we're ready to clip the red and infrared bands using the bounding box:

```
# Clip red and infrared to new bounds.
rClip = r[ulY:lrY, ulX:lrX]
irClip = ir[ulY:lrY, ulX:lrX]
```

7. Next, we create the georeferencing information for the image:

```
# Create a new geomatrix for the image
geoTrans = list(geoTrans)
geoTrans[0] = minX
geoTrans[3] = maxY
```

8. Then we can prepare to map points to pixels in order to create our mask image:

```
# Map points to pixels for drawing
# the field boundary on a blank
# 8-bit, black and white, mask image.
points = []
pixels = []
# Grab the polygon geometry
geom = poly.GetGeometryRef()
pts = geom.GetGeometryRef(0)
```

9. We loop through all of the point features and store their *x* and *y* values:

```
# Loop through geometry and turn
# the points into an easy-to-manage
# Python list
for p in range(pts.GetPointCount()):
    points.append((pts.GetX(p), pts.GetY(p)))
```

10. Now, we convert the points to pixel locations:

```
# Loop through the points and map to pixels.
# Append the pixels to a pixel list
for p in points:
    pixels.append(world2Pixel(geoTrans, p[0], p[1]))
```

11. Next, we create a new image that will serve as our mask image:

```
# Create the raster polygon image as a black and white 'L' mode
# and filled as white. White=1
rasterPoly = Image.new("L", (pxWidth, pxHeight), 1)
```

12. Now we can rasterize our polygon:

```
# Create a PIL drawing object
rasterize = ImageDraw.Draw(rasterPoly)

# Dump the pixels to the image
# as a polygon. Black=0
rasterize.polygon(pixels, 0)
```

13. Finally, we can convert our mask to a numpy array:

```
# Hand the image back to gdal/gdal_array
# so we can use it as an array mask
mask = imageToArray(rasterPoly)
```

Now that we have converted the shapefile to a mask image, we can clip the bands.

Clipping the bands

Now that we have our image mask, we can clip the red and infrared bands to the boundary of the mask. For this process, we use NumPy's choose() method that correlates the mask cell to the raster band cell and returns that value, or returns 0. The result is a new array that is clipped to the mask, but with the correlated values from the raster band:

```
# Clip the red band using the mask
rClip = gdal_array.numpy.choose(mask,
  (rClip, 0)).astype(gdal_array.numpy.uint8)

# Clip the infrared band using the mask
irClip = gdal_array.numpy.choose(mask,
  (irClip, 0)).astype(gdal_array.numpy.uint8)
```

We now have just the data that we want, so we can apply our NDVI relative vegetation health formula.

Using the NDVI formula

Our final process for creating the NDVI is to execute the equation that is *infrared - red/infrared + red*. The first step that we perform silences any **not-a-number**, also known as **NaN**, values in NumPy that might occur during division. And before we save the output, we'll convert any NaN values to 0. We'll save the output as `ndvi.tif`, and that will be the input for the next script in order to classify and colorize the NDVI as follows:

1. First, we'll ignore any warnings from `numpy`, as we'll get some errors near the edges:

   ```
   # We don't care about numpy warnings
   # due to NaN values from clipping
   gdal_array.numpy.seterr(all="ignore")
   ```

2. Now we can perform our NDVI formula:

   ```
   # NDVI equation: (infrared - red) / (infrared + red)
   # *1.0 converts values to floats,
   # +1.0 prevents ZeroDivisionErrors
   ndvi = 1.0 * ((irClip - rClip) / (irClip + rClip + 1.0))
   ```

3. If there are any NaN values, we convert them to zero:

   ```
   # Convert any NaN values to 0 from the final product
   ndvi = gdal_array.numpy.nan_to_num(ndvi)
   ```

4. Finally, we save our finished NDVI image:

   ```
   # Save the ndvi as a GeoTIFF and copy/adjust
   # the georeferencing info
   gtiff = gdal.GetDriverByName( 'GTiff' )
   gtiff.CreateCopy(target, copy_geo(ndvi, prototype=source,
   xoffset=ulX, yoffset=ulY))
   gtiff = None
   ```

The following figure is the output of this example. You need to view it in a geospatial viewer such as QGIS or OpenEV. The image won't open in most image editors. The lighter the shade of gray, the healthier the plant is within that field:

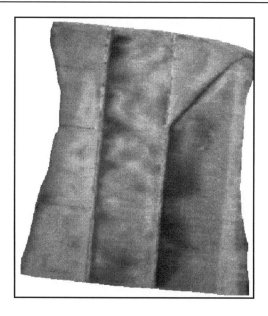

Now that we know how to use the NDVI formula, let's see how to classify it.

Classifying the NDVI

We now have a valid index, but it is not easy to understand, because it is a grayscale image. If we color the image in an intuitive way, then even a child can identify the healthier plants. In the following section, *Additional functions*, we read in this grayscale index and classify it from brown to dark green using seven classes. The classification and image processing routines, such as the histogram and stretching functions, are almost identical to what we used in the *Creating histograms* section in Chapter 6, *Python and Remote Sensing*, but this time we are applying them in a much more specific way.

The output of this example will be another GeoTIFF file, but this time it will be a colorful RGB image.

Additional functions

We won't need any of the functions from our previous NDVI script, but we do need to add a function for creating and stretching a histogram. Both of these functions work with NumPy arrays. We'll also shorten the reference to `gdal_array` to `gd` in this script because it is a long name, and we need it throughout the script.

Let's have a look at the steps as follows:

1. First, we import the libraries that we need:

```
import gdal_array as gd
import operator
from functools import reduce
```

2. Next, we need to create a `histogram` function, which we'll need in order to do a histogram stretch:

```
def histogram(a, bins=list(range(256))):
    """
    Histogram function for multi-dimensional array.
    a = array
    bins = range of numbers to match
    """
    # Flatten, sort, then split our arrays for the histogram.
    fa = a.flat
    n = gd.numpy.searchsorted(gd.numpy.sort(fa), bins)
    n = gd.numpy.concatenate([n, [len(fa)]])
    hist = n[1:]-n[:-1]
    return hist
```

3. Now, we create our histogram `stretch` function:

```
def stretch(a):
    """
    Performs a histogram stretch on a gdal_array array image.
    """
    hist = histogram(a)
    lut = []
    for b in range(0, len(hist), 256):
    # step size - create equal interval bins.
    step = reduce(operator.add, hist[b:b+256]) / 255
    # create equalization lookup table
    n = 0
    for i in range(256):
    lut.append(n / step)
    n = n + hist[i+b]
    gd.numpy.take(lut, a, out=a)
    return a
```

Now that we have our utility functions, we can process the NDVI.

Loading the NDVI

Next, we'll load the output of our NDVI script back into a NumPy array. We'll also define the name of our output image as `ndvi_color.tif`, and create a zero-filled multidimensional array as a placeholder for the red, green, and blue bands of the colorized NDVI image. The following code will load the NDVI TIFF image into a `numpy` array:

```
# NDVI output from ndvi script
source = "ndvi.tif"

# Target file name for classified
# image image
target = "ndvi_color.tif"

# Load the image into an array
ndvi = gd.LoadFile(source).astype(gd.numpy.uint8)
```

Now that our image is loaded as an array, we can stretch it.

Preparing the NDVI

We need to perform a histogram stretch on the NDVI in order to ensure that the image covers the range of classes that will give the final product meaning:

```
# Peform a histogram stretch so we are able to
# use all of the classes
ndvi = stretch(ndvi)

# Create a blank 3-band image the same size as the ndvi
rgb = gd.numpy.zeros((3, len(ndvi), len(ndvi[0])), gd.numpy.uint8)
```

Now that we've stretched the image, we can begin the classification process.

Creating classes

In this part, we set up the ranges for our NDVI classes, which are broken up across a range from 0 to 255. We'll use seven classes. You can change the number of classes by adding or removing values from the classes list. Next, we create a **look-up table**, or **LUT**, in order to assign colors for each class. The number of colors must match the number of classes.

The colors are defined as RGB values. The `start` variable defines the beginning of the first class. In this case, 0 is a nodata value, which we designated in the previous script, so we begin the class at 1. We then loop through the classes, extract the ranges, and use the color assignments to add the RGB value to our placeholder array. Finally, we save the colorized image as a GeoTIFF file:

```
# Class list with ndvi upper range values.
# Note the lower and upper values are listed on the ends
classes = [58, 73, 110, 147, 184, 220, 255]

# Color look-up table (lut)
# The lut must match the number of classes
# Specified as R, G, B tuples from dark brown to dark green
lut = [[120, 69, 25], [255, 178, 74], [255, 237, 166], [173, 232, 94],
    [135, 181, 64], [3, 156, 0], [1, 100, 0]]

# Starting value of the first class
start = 1
```

Now we can classify the image:

```
# For each class value range, grab values within range,
# then filter values through the mask.
for i in range(len(classes)):
 mask = gd.numpy.logical_and(start <= ndvi,
 ndvi <= classes[i])
 for j in range(len(lut[i])):
     rgb[j] = gd.numpy.choose(mask, (rgb[j], lut[i][j]))
     start = classes[i]+1
```

Finally, we can save our classified GeoTIFF file:

```
# Save a geotiff image of the colorized ndvi.
output=gd.SaveArray(rgb, target, format="GTiff", prototype=source)
output = None
```

Here is the image that we output:

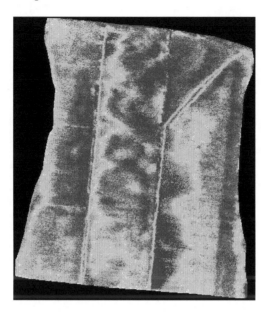

This is our final product for this example. Farmers can use this data to determine how to effectively irrigate and spray chemicals, such as fertilizers and pesticides, in a targeted, more effective, and more environmentally friendly way. In fact, these classes can even be turned into a vector shapefile, which is then loaded into a GPS-driven computer on a field sprayer. This then automatically applies the correct amount of chemicals in the correct place as a sprayer is driven around the field, or in some cases, even flown over the field in an airplane with a sprayer attachment.

Notice as well that even though we clipped the data to the field, the image is still a square. The black areas are the nodata values that have been converted to black. In display software, you can make the nodata color transparent without affecting the rest of the image.

Although we created a very specific type of product, a classified NDVI, the framework of this script can be altered in order to implement many remote sensing analysis algorithms. There are different types of NDVIs, but with relatively minor changes, you can turn this script into a tool that can be used to look for harmful algae blooms in the ocean, or smoke in the middle of a forest indicating a forest fire.

This book attempts to limit the use of GDAL as much as possible in order to focus on what can be accomplished with pure Python and tools that can easily be installed from PyPI. However, it is helpful to remember that there is a wealth of information on using GDAL and its associated utilities to carry out similar tasks. For another tutorial on clipping a raster with GDAL via its command-line utilities, see `https://joeyklee.github.io/broc-cli-geo/guide/XX_raster_cropping_and_clipping.html`.

Now that we've worked with the land, let's work with water in order to create a flood inundation model.

Creating a flood inundation model

In this next example, we'll begin to enter the world of hydrology. Flooding is one of the most common and devastating natural disasters, which affects nearly every population on the globe. Geospatial models are a powerful tool in estimating the impact of a flood and mitigating that impact before it happens. We often hear on the news that a river is reaching the flood stage, but that information is meaningless if we can't understand the impact.

Hydrological flood models are expensive to develop and can be very complex. These models are essential for engineers in building flood control systems. However, first responders and potential flood victims are only interested in the impact of an impending flood.

We can begin to understand the flooding impact in an area using a very simple and easy-to-comprehend tool called a **flood inundation model**. This model starts with a single point and floods an area with the maximum volume of water that a flood basin can hold at a particular flood stage. Usually, this analysis is a worst-case scenario. Hundreds of other factors go into calculating how much water will enter into a basin from a river-topping flood stage. But we can still learn a lot from this simple first-order model.

As mentioned in the *Elevation data* section in Chapter 1, *Learning about Geospatial Analysis with Python*, the **Shuttle Radar Topography Mission (SRTM)** dataset provides a nearly-global DEM that you can use for these types of models. More on SRTM data can be found here: `http://www2.jpl.nasa.gov/srtm/`.

You can download the ASCII Grid data in EPSG:4326, and a shapefile containing the point as a .zip file from http://git.io/v3fSg. The shapefile is just for reference and has no role in this model. The following image is a **digital elevation model** (**DEM**) with a source point displayed as a yellow star near Houston, Texas. In real-world analysis, this point would likely be a stream gauge where you would have data about the river's water level:

The algorithm that we are introducing in this example is called a **flood fill algorithm**. This algorithm is well known in the field of computer science and is used in the classic computer game **Minesweeper** to clear empty squares on the board when a user clicks a square. It is also the method that is used for the well-known **paint bucket tool** in graphics programs such as **Adobe Photoshop**, and it is used to fill an area of adjacent pixels of the same color with a different color.

There are many ways to implement this algorithm. One of the oldest and most common ways is to recursively crawl through each pixel of the image. The problem with recursion is that you end up processing pixels more than once and creating an unnecessary amount of work. The resource usage for a recursive flood fill can easily crash a program on even a moderately sized image.

This script uses a four-way queue-based flood fill that may visit a cell more than once but ensures that we only process a cell once. The queue only contains unique, unprocessed cells by using Python's built-in set type, which only holds unique values. We use two sets: **fill**, which contains the cells we need to fill, and **filled**, which contains processed cells.

This example executes the following steps:

1. Extract the header information from the ASCII DEM.
2. Open the DEM as a `numpy` array.
3. Define our starting point as row and column in the array.
4. Declare a flood elevation value.
5. Filter the terrain to only the desired elevation value and below.
6. Process the filtered array.
7. Create a 1, 0, 0 array (that is, a binary array) with flooded pixels as 1.
8. Save the flood inundation array as an ASCII Grid.

 This example can take a minute or two to run on a slower machine; we'll use the `print` statements throughout the script as a simple way to track progress. Once again we'll break this script up with explanations, for clarity.

Now that we have our data, we can begin our flood fill function.

The flood fill function

We use ASCII Grids in this example, which means that the engine for this model is completely in NumPy. We start off by defining the `floodFill()` function, which is the heart and soul of this model. This Wikipedia article on flood fill algorithms provides an excellent overview of the different approaches: `http://en.wikipedia.org/wiki/Flood_fill`.

Flood fill algorithms start at a given cell and begin checking the neighboring cells for similarity. The similarity factor might be color or, in our case, elevation. If the neighboring cell is of the same or lower elevation as the current cell, then that cell is marked for checks of its neighbor until the entire grid is checked. NumPy isn't designed to crawl over an array in this way, but it is still efficient in handling multidimensional arrays overall. We step through each cell and check its neighbors to the north, south, east, and west. Any of those cells which can be flooded are added to the filled set, and their neighbors are added to the fill set to be checked by the algorithm.

As mentioned earlier, if you try to add the same value to a set twice, it just ignores the duplicate entry and maintains a unique list. By using sets in an array, we efficiently check a cell only once because the fill set contains unique cells. The following code implements our `floodFill` function:

1. First we import our libraries:

```
import numpy as np
from linecache import getline
```

2. Next, we create our `floodFill` function:

```
def floodFill(c, r, mask):
    """
    Crawls a mask array containing
    only 1 and 0 values from the
    starting point (c=column,
    r=row - a.k.a. x, y) and returns
    an array with all 1 values
    connected to the starting cell.
    This algorithm performs a 4-way
    check non-recursively.
    """
```

3. Next, we create sets to track the cells that we've already covered:

```
# cells already filled
filled = set()
# cells to fill
fill = set()
fill.add((c, r))
width = mask.shape[1]-1
height = mask.shape[0]-1
```

4. Then we create our inundation array:

```
# Our output inundation array
flood = np.zeros_like(mask, dtype=np.int8)
```

5. Now we can loop through the cells and flood them, or not:

```
# Loop through and modify the cells which
# need to be checked.
while fill:
    # Grab a cell
    x, y = fill.pop()
```

6. If the land is higher than the floodwater, skip it:

```
if y == height or x == width or x < 0 or y < 0:
 # Don't fill
 continue
```

7. If the land elevation is equal to or less than the floodwater, fill it in:

```
if mask[y][x] == 1:
 # Do fill
 flood[y][x] = 1
filled.add((x, y))
```

8. Now, we check the surrounding neighbor cells to see if they need to be filled, and when we run out of cells, we return the flooded matrix:

```
# Check neighbors for 1 values
west = (x-1, y)
east = (x+1, y)
north = (x, y-1)
south = (x, y+1)
if west not in filled:
  fill.add(west)
if east not in filled:
  fill.add(east)
if north not in filled:
  fill.add(north)
if south not in filled:
  fill.add(south)
return flood
```

Now that we've set up our `floodFill` function, we can create a flood.

Predicting flood inundation

In the remainder of the script, we load our terrain data from an ASCII Grid, define our output grid filename, and execute the algorithm on the terrain data. The seed of the flood fill algorithm is an arbitrary point, as `sx` and `sy` within the lower elevation areas. In a real-world application, these points would likely be a known location, such as a stream gauge or a breach in a dam. In the final step, we save the output grid.

The following steps need to be performed:

1. First, we set up our `source` and `target` data names:

   ```
   source = "terrain.asc"
   target = "flood.asc"
   ```

2. Next, we open the source:

   ```
   print("Opening image...")
   img = np.loadtxt(source, skiprows=6)
   print("Image opened")
   ```

3. We'll create a mask array of everything below 70 meters:

   ```
   # Mask elevations lower than 70 meters.
   wet = np.where(img < 70, 1, 0)
   print("Image masked")
   ```

4. Now, we'll parse the geospatial information from the header:

   ```
   # Parse the header using a loop and
   # the built-in linecache module
   hdr = [getline(source, i) for i in range(1, 7)]
   values = [float(h.split(" ")[-1].strip()) for h in hdr]
   cols, rows, lx, ly, cell, nd = values
   xres = cell
   yres = cell * -1
   ```

5. Now, we'll establish a starting point that is located in a riverbed:

   ```
   # Starting point for the
   # flood inundation in pixel coordinates
   sx = 2582
   sy = 2057
   ```

6. Now, we trigger our `floodFill` function:

   ```
   print("Beginning flood fill")
   fld = floodFill(sx, sy, wet)
   print("Finished flood fill")

   header = ""
   for i in range(6):
    header += hdr[i]
   ```

7. Finally, we can save our flood inundation model output:

```
print("Saving grid")
# Open the output file, add the hdr, save the array
with open(target, "wb") as f:
  f.write(bytes(header, 'UTF-8'))
  np.savetxt(f, fld, fmt="%1i")
print("Done!")
```

The image in the following screenshot shows the flood inundation output over a classified version of the DEM, with lower elevation values in brown, mid-range values in green, and higher values in gray and white:

The flood raster, which includes all areas less than 70 meters, is colored blue. This image was created with QGIS, but it could be displayed in ArcGIS as EPSG:4326. You could also use GDAL to save the flood raster grid as an 8-bit TIFF file or JPEG file, just like the NDVI example, in order to view it in a standard graphics program.

This image in the following screenshot is nearly identical, except for the filtered mask from which the inundation was derived, which is displayed in yellow. This is done by generating a file for the array called `wet`, instead of `fld`, to show the non-contiguous regions, which were not included as part of a flood. These areas are not connected to the source point, so they would unlikely be reached during a flood event:

By changing the elevation value, you can create additional flood inundation rasters. We started with an elevation of 70 meters. If we increase that value to 90, we can expand the flood. The following screenshot shows a flood event at both 70 and 90 meters:

The 90 meter inundation is the lighter-blue polygon. You can take bigger or smaller steps and show different impacts as different layers.

This model is an excellent and useful visualization. However, you could take this analysis even further by using GDAL's `polygonize()` method on the flood mask, as we did with the island in the *Extracting features from images* section in Chapter 6, *Python and Remote Sensing*. This operation would give you a vector flood polygon. Then, you could use the principles that we discussed in the *Performing selections* section in Chapter 5, *Python and Geographic Information Systems*, to select buildings using the polygon to determine population impact. You could also combine that flood polygon with the dot density example in Chapter 5, *Python and Geographic Information Systems*, in the *Dot density calculations* section, to assess the potential population impact of a flood. The possibilities are endless.

Creating a color hillshade

In this example, we'll combine previous techniques to combine our terrain hillshade from Chapter 7, *Python and Elevation Data*, with the color classification that we used on the LIDAR. For this example, we'll need the ASCII Grid DEMs named `dem.asc` and `relief.asc` that we used in the previous chapter.

We'll create a colorized DEM and a hillshade, and then use PIL to blend them together for an enhanced elevation visualization. The code comments will guide you through the example, as many of these steps are already familiar to you:

1. First, we import the libraries that we need:

```
import gdal_array as gd
try:
  import Image
except ImportError:
  from PIL import Image
```

   For this next part, you'll need the following two files: https://github.com/GeospatialPython/Learn/raw/master/relief.zip and https://github.com/GeospatialPython/Learn/raw/master/dem.zip.

2. Then, we'll set up variables for the inputs and outputs:

```
relief = "relief.asc"
dem = "dem.asc"
target = "hillshade.tif"
```

3. Next, we'll load our `relief` image:

```
# Load the relief as the background image
bg = gd.numpy.loadtxt(relief, skiprows=6)
```

4. Then, we'll load the DEM image, so that we'll have the elevation data:

```
# Load the DEM into a numpy array as the foreground image
fg = gd.numpy.loadtxt(dem, skiprows=6)[:-2, :-2]
```

5. Now, we'll create a new image for our colorization with elevation breakpoints forming classes and corresponding colors in a LUT:

```
# Create a blank 3-band image to colorize the DEM
rgb = gd.numpy.zeros((3, len(fg), len(fg[0])), gd.numpy.uint8)

# Class list with DEM upper elevation range values.
classes = [356, 649, 942, 1235, 1528,
 1821, 2114, 2300, 2700]

# Color look-up table (lut)
# The lut must match the number of classes.
# Specified as R, G, B tuples
lut = [[63, 159, 152], [96, 235, 155], [100, 246, 174],
 [248, 251, 155], [246, 190, 39], [242, 155, 39],
 [165, 84, 26], [236, 119, 83], [203, 203, 203]]

# Starting elevation value of the first class
start = 1
```

6. We can now perform our color classification:

```
# Process all classes.
for i in range(len(classes)):
 mask = gd.numpy.logical_and(start <= fg,
 fg <= classes[i])
 for j in range(len(lut[i])):
 rgb[j] = gd.numpy.choose(mask, (rgb[j], lut[i][j]))
 start = classes[i]+1
```

7. Then, we can convert our shaded relief array to an image, as well as our colorized DEM:

```
# Convert the shaded relief to a PIL image
im1 = Image.fromarray(bg).convert('RGB')

# Convert the colorized DEM to a PIL image.
# We must transpose it from the Numpy row, col order
# to the PIL col, row order (width, height).
im2 = Image.fromarray(rgb.transpose(1, 2, 0)).convert('RGB')
```

8. Now, we'll blend the two images for the final effect and save it to an image file:

```
# Blend the two images with a 40% alpha
hillshade = Image.blend(im1, im2, .4)

# Save the hillshade
hillshade.save(target)
```

The following image shows the output, which makes a great backdrop for GIS maps:

Now that we can model terrain, let's learn how to navigate over it.

Performing least cost path analysis

Calculating driving directions is the most commonly used geospatial function in the world. Typically, these algorithms calculate the shortest path between points *A* and *B*, or they may take into account the speed limit of the road, or even current traffic conditions, in order to choose a route by drive time.

But what if your job is to build a new road? Or what if you are in charge of deciding where to run power transmission lines or water lines across a remote area? In a terrain-based setting, the shortest path might cross a difficult mountain, or run through a lake. In this case, we need to account for obstacles and avoid them if possible. However, if avoiding a minor obstacle takes us too far out of our way, the cost of implementing that route may be more expensive than just going over a mountain.

This type of advanced analysis is called **least cost path analysis**. We search an area for the route that is the best compromise of distance versus the cost of following that route. The algorithm that we use for this process is called the **A-star or A*** algorithm. The oldest routing method is called the **Dijkstra algorithm**, which calculates the shortest path in a network, such as a road network. The A* method can do that as well, but it is also better suited for traversing a grid-like DEM.

You can find out more about these algorithms on the following web pages:

- Dijkstra's algorithm: `http://en.wikipedia.org/wiki/ Dijkstra's_algorithm`.
- A* algorithm: `http://en.wikipedia.org/wiki/A-star_ algorithm`.

This example is the most complex in this chapter. To better understand it, we have a simple version of the program, which is text based, and operates on a 5 x 5 grid with randomly generated values. You can actually see how this program follows the algorithm before trying it on an elevation grid with thousands of values.

This program executes the following steps:

1. Create a simple grid with randomly generated pseudo-elevation values between 1 and 16.
2. Define a start location in the lower-left corner of the grid.
3. Define the end point as the upper-right corner of the grid.

4. Create a cost grid that has the elevation of each cell, plus the cell's distance to the finish.

5. Examine each neighboring cell from the start, and choose the one with the lowest cost.

6. Repeat the evaluation using the chosen cell until we get to the end.

7. Return the set of chosen cells as the least cost path.

8. Set up the test grid.

You simply run this program from the command line and view its output. The first section of this script sets up our artificial terrain grid as a randomly generated NumPy array, with notional elevation values between 1 and 16. We also create a distance grid that calculates the distance for each cell to the destination cell. This value is the cost of each cell.

Let's have a look at the following steps:

1. First, we'll import numpy and set the size of our grid:

```
import numpy as np

# Width and height
# of grids
w = 5
h = 5
```

2. Next, we set a starting location cell and an ending location:

```
# Start location:
# Lower left of grid
start = (h-1, 0)

# End location:
# Top right of grid
dx = w-1
dy = 0
```

3. Now, we can create a grid of zeros based on our width and height:

```
# Blank grid
blank = np.zeros((w, h))
```

4. Next, we'll set up our distance grid in order to create impedance values:

```
# Distance grid
dist = np.zeros(blank.shape, dtype=np.int8)

# Calculate distance for all cells
for y, x in np.ndindex(blank.shape):
  dist[y][x] = abs((dx-x)+(dy-y))
```

5. Now, we'll print out the cost value of each cell in our cost grid:

```
# "Terrain" is a random value between 1-16.
# Add to the distance grid to calculate
# The cost of moving to a cell
cost = np.random.randint(1, 16, (w, h)) + dist

print("COST GRID (Value + Distance)\n{}\n".format(cost))
```

Now that we have a simulated terrain grid to work with, we can test a routing algorithm.

The simple A* algorithm

The A* search algorithm that is implemented here crawls the grid in a similar fashion to our flood fill algorithm in the previous example. Once again, we use sets to avoid using recursion, and to avoid the duplication of cell checks. But this time, instead of checking elevation, we check the distance cost of routing through a cell in question. If the move raises the cost of getting to the end, then we go with a lower-cost option.

The following steps need to be performed, as follows:

1. First, we'll start our A* function by creating sets that will keep track of the path progress:

```
# Our A* search algorithm
def astar(start, end, h, g):
    closed_set = set()
    open_set = set()
    path = set()
```

2. Next, we add the starting cell to the open list of cells in order to process and begin looping through that set:

```
open_set.add(start)
while open_set:
    cur = open_set.pop()
    if cur == end:
        return path
    closed_set.add(cur)
    path.add(cur)
    options = []
    y1 = cur[0]
    x1 = cur[1]
```

3. We check the surrounding cells as options for forward progress:

```
if y1 > 0:
    options.append((y1-1, x1))
if y1 < h.shape[0]-1:
    options.append((y1+1, x1))
if x1 > 0:
    options.append((y1, x1-1))
if x1 < h.shape[1]-1:
    options.append((y1, x1+1))
if end in options:
    return path
best = options[0]
closed_set.add(options[0])
```

4. We then check each option for the best option and append it to the path until we reach the end:

```
for i in range(1, len(options)):
    option = options[i]
    if option in closed_set:
        continue
    elif h[option] <= h[best]:
        best = option
        closed_set.add(option)
    elif g[option] < g[best]:
        best = option
        closed_set.add(option)
    else:
        closed_set.add(option)
print(best, ", ", h[best], ", ", g[best])
open_set.add(best)
return []
```

Now that we have the algorithm set up, we can test it out by creating a path.

Generating the test path

In this section, we'll generate a path on our test grid. We'll call our A* function, using the starting point, end point, cost grid, and distance grid:

```
# Find the path
path = astar(start, (dy, dx), cost, dist)
print()
```

Now, we'll put our path on its own grid and print it:

```
# Create and populate the path grid
path_grid = np.zeros(cost.shape, dtype=np.uint8)
for y, x in path:
 path_grid[y][x] = 1
path_grid[dy][dx] = 1

print("PATH GRID: 1=path")
print(path_grid)
```

Next, we'll view the output of this test.

Viewing the test output

When you run this program, you'll generate a randomly-numbered grid similar to the following:

```
COST GRID (Value + Distance)
[[13 10 5 15 9]
 [15 13 16 5 16]
 [17 8 9 9 17]
 [ 4 1 11 6 12]
 [ 2 7 7 11 8]]

(Y,X), HEURISTIC, DISTANCE
(3, 0) , 4 , 1
(3, 1) , 1 , 0
(2, 1) , 8 , 1
(2, 2) , 9 , 0
(2, 3) , 9 , 1
(1, 3) , 5 , 0
(0, 3) , 15 , 1
```

```
PATH GRID: 1=path
[[0 0 0 1 1]
 [0 0 0 1 0]
 [0 1 1 1 0]
 [1 1 0 0 0]
 [1 0 0 0 0]]
```

The grid is small enough such that you can easily trace the algorithm's steps manually. This implementation uses **Manhattan distance**, which means the distance does not use diagonal lines—only left, right, up, and down measurements. The search also does not move diagonally in order to keep things simple.

The real-world example

Now that we have a basic understanding of the A* algorithm, let's move to a more complex example. For the relief example, we'll use the same DEM that is located near Vancouver, British Columbia, Canada, which we used in Chapter 7, *Python and Elevation Data*, in the *Creating a shaded relief* section. The spatial reference for this grid is EPSG:26910 NAD 83/UTM zone 10N. You can download the DEM, relief, and start and end points of the shapefile as a zipped package from http://git.io/v3fpL.

We'll actually use the shaded relief for visualization. Our goal in this exercise will be to move from the start to the finish point in the lowest-cost way possible:

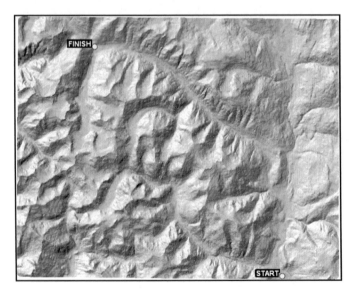

Just looking at the terrain, there are two paths that follow low-elevation routes without much change in direction. These two routes are illustrated in the following screenshot:

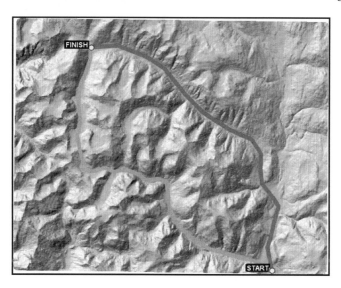

So, we would expect that when we used the A* algorithm, it would be close. Remember that the algorithm is only looking in the immediate vicinity, so it can't look at the whole image like we can, and it can't make adjustments early in the route based on a known obstacle ahead.

We will expand this implementation from our simple example and use Euclidean distance, or *as the crow flies* measurements, and we will also allow the search to look in eight directions instead of four. We will prioritize terrain as the primary decision point. We will also use distance, both to the finish and from the start, as lower priorities in order to make sure that we are moving forward toward the goal and not getting too far off track. Other than those differences, the steps are identical to the simple example. The output will be a raster with the path values set to one and the other values set to zero.

Now that we understand the problem, let's solve it!

Loading the grid

In this section and the following sections, we'll create the script that can create a route over terrain. The script starts out simple enough. We load the grid into a NumPy array from an ASCII Grid. We name our output path grid, and then we define the starting cell and end cell:

1. First, we import our libraries:

```
import numpy as np
import math
from linecache import getline
import pickle
```

2. Next, we'll define our input and output data sources:

```
# Our terrain data
source = "dem.asc"

# Output file name for the path raster
target = "path.asc"
```

3. Then, we can load the grid skipping over the header:

```
print("Opening %s..." % source)
cost = np.loadtxt(source, skiprows=6)
print("Opened %s." % source)
```

4. Next, we'll parse the header for the geospatial and grid size information:

```
# Parse the header
hdr = [getline(source, i) for i in range(1, 7)]
values = [float(ln.split(" ")[-1].strip()) for ln in hdr]
cols, rows, lx, ly, cell, nd = values
```

5. Finally, we'll define our starting and end locations:

```
# Starting column, row
sx = 1006
sy = 954

# Ending column, row
dx = 303
dy = 109
```

Now that our grid is loaded, we can set up the functions that we'll need.

Defining the helper functions

We need three functions in order to route over terrain. One is the A* algorithm, and the other two assist the algorithm in choosing the next step. We'll briefly discuss these helper functions. First, we have a simple Euclidean distance function named e_dist, which returns the straight-line distance between two points as map units. Next, we have an important function called `weighted_score`, which returns a score for a neighboring cell, based on the elevation change between the neighbor and the current cell, as well as the distance to the destination.

This function is better than distance or elevation alone because it reduces the chance of there being a tie between two cells, making it easier to avoid back-tracking. This scoring formula is loosely based on a concept called the **Nisson Score**, which is commonly used in these types of algorithms and is referenced in the Wikipedia articles mentioned earlier in this chapter. What's great about this function is that it can score the neighboring cell with any values that you wish. You might also use a real-time feed to look at the current weather in the neighboring cell, and avoid cells with rain or snow.

The following code will create our distance function and our weighting function that we'll need to traverse the terrain:

1. First, we'll create a Euclidean distance function that will give us the distance between points:

```
def e_dist(p1, p2):
    """
    Takes two points and returns
    the Euclidian distance
    """
    x1, y1 = p1
    x2, y2 = p2
    distance = math.sqrt((x1-x2)**2+(y1-y2)**2)
    return int(distance)
```

2. Now, we'll create our weight function in order to score each node for its suitability to move:

```
def weighted_score(cur, node, h, start, end):
    """
    Provides a weighted score by comparing the
    current node with a neighboring node. Loosely
    based on the Nisson Score concept: f=g+h
    In this case, the "h" value, or "heuristic",
    is the elevation value of each node.
    """
```

3. We start with a `score` of 0 and check the node's distance from the end and the start:

```
score = 0
# current node elevation
cur_h = h[cur]
# current node distance from end
cur_g = e_dist(cur, end)
# current node distance from
cur_d = e_dist(cur, start)
```

4. Next, we examine the neighboring nodes and make a decision on where to move:

```
# neighbor node elevation
node_h = h[node]
# neighbor node distance from end
node_g = e_dist(node, end)
# neighbor node distance from start
node_d = e_dist(node, start)
# Compare values with the highest
# weight given to terrain followed
# by progress towards the goal.
if node_h < cur_h:
score += cur_h-node_h
if node_g < cur_g:
score += 10
if node_d > cur_d:
score += 10
return score
```

Now that our helper functions are complete, we can build the A* function.

The real-world A* algorithm

This algorithm is more involved than the simple version in our previous example. We use sets to avoid redundancy. It also implements our more advanced scoring algorithm and checks to make sure we aren't at the end of the path before doing additional calculations. Unlike our last example, this more advanced version also checks cells in eight directions, so the path can move diagonally. There is a `print` statement at the end of this function that is commented out. You can uncomment it in order to watch the search crawl through the grid. The following code will implement the A* algorithm that we will use for the rest of the section:

1. First, we open the function by accepting a starting point, an end point, and a score:

```
def astar(start, end, h):
    """
    A-Star (or A*) search algorithm.
    Moves through nodes in a network
    (or grid), scores each node's
    neighbors, and goes to the node
    with the best score until it finds
    the end. A* is an evolved Dijkstra
    algorithm.
    """
```

2. Now, we set up the sets that will track progress:

```
# Closed set of nodes to avoid
closed_set = set()
# Open set of nodes to evaluate
open_set = set()
# Output set of path nodes
path = set()
```

3. Next, we begin processing using our starting point:

```
# Add the starting point to
# to begin processing
open_set.add(start)
while open_set:
# Grab the next node
cur = open_set.pop()
```

4. If we hit the end, we return the completed path:

```
# Return if we're at the end
if cur == end:
return path
```

5. Otherwise, we keep working through the grid and eliminating possibilities:

```
# Close off this node to future
# processing
closed_set.add(cur)
# The current node is always
# a path node by definition
path.add(cur)
```

6. To keep things moving, we grab all of the neighbors that need to be processed as we go:

```
# List to hold neighboring
# nodes for processing
options = []
# Grab all of the neighbors
y1 = cur[0]
x1 = cur[1]
if y1 > 0:
options.append((y1-1, x1))
if y1 < h.shape[0]-1:
options.append((y1+1, x1))
if x1 > 0:
options.append((y1, x1-1))
if x1 < h.shape[1]-1:
options.append((y1, x1+1))
if x1 > 0 and y1 > 0:
options.append((y1-1, x1-1))
if y1 < h.shape[0]-1 and x1 < h.shape[1]-1:
options.append((y1+1, x1+1))
if y1 < h.shape[0]-1 and x1 > 0:
options.append((y1+1, x1-1))
if y1 > 0 and x1 < h.shape[1]-1:
options.append((y1-1, x1+1))
```

7. We check each neighbor for being the destination:

```
# If the end is a neighbor, return
if end in options:
return path
```

8. We take the first option as the `best` option and process the other options, upgrading as we go:

```
# Store the best known node
best = options[0]
# Begin scoring neighbors
best_score = weighted_score(cur, best, h, start, end)
# process the other 7 neighbors
for i in range(1, len(options)):
option = options[i]
# Make sure the node is new
if option in closed_set:
continue
else:
# Score the option and compare
# it to the best known
option_score = weighted_score(cur, option,
h, start, end)
if option_score > best_score:
best = option
best_score = option_score
else:
# If the node isn't better seal it off
closed_set.add(option)
# Uncomment this print statement to watch
# the path develop in real time:
# print(best, e_dist(best, end))
# Add the best node to the open set
open_set.add(best)
return []
```

Now that we have our routing algorithm, we can generate a real-world path.

Generating a real-world path

Finally, we create our real-world path as a chain of ones in a grid of zeros. This raster can then be brought into an application such as QGIS and visualized over the terrain grid. In the following code, we'll use our algorithm and helper functions to generate a path, as follows:

1. First, we send our start and end points, as well as our terrain grid, to the routing function:

```
print("Searching for path...")
p = astar((sy, sx), (dy, dx), cost)
```

```
print("Path found.")
print("Creating path grid...")
path = np.zeros(cost.shape)
print("Plotting path...")
for y, x in p:
 path[y][x] = 1
path[dy][dx] = 1
print("Path plotted.")
```

2. Once we have a path, we can save it out as an ASCII Grid:

```
print("Saving %s..." % target)
header = ""
for i in range(6):
 header += hdr[i]

# Open the output file, add the hdr, save the array
with open(target, "wb") as f:
 f.write(bytes(header, 'UTF-8'))
 np.savetxt(f, path, fmt="%4i")
```

3. Now, we want to save our path data because the points are in the correct order, from the starting point to the end point. When we put them into the grid, we lose that order because it is all one raster. We'll use the built-in Python `pickle` module to save the list object to disk. We're going to use this data in the next section to create a vector shapefile of the route. So, we'll save our path data as a pickled Python object that we can reuse later, without running the whole program:

```
print("Saving path data...")
with open("path.p", "wb") as pathFile:
 pickle.dump(p, pathFile)
print("Done!")
```

Here is the output route of our search:

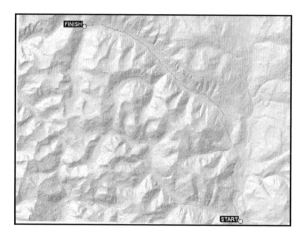

As you can see, the A* search came very close to one of our manually selected routes. In a couple of cases, the algorithm chose to tackle some terrain, instead of trying to go around it. Sometimes the slight terrain is deemed less of a cost than the distance to go around it. You can see examples of that choice in this zoomed-in portion of the upper-right section of the route. The red line is the route that our program generated through the terrain:

We only used two values: terrain and distance. But you could also add hundreds of factors, such as soil type, water bodies, and existing roads. All of these items could serve as an impedance or an outright wall. You would just modify the scoring function in the example to account for any additional factors. Keep in mind, the more factors you add, the more difficult it is to trace what the A* implementation was *thinking* when it chose the route.

An obvious future direction for this analysis would be to create a vector version of this route as a line. The process would include mapping each cell to a point and then using nearest-neighbor analysis to order the points properly, before saving it as a shapefile or GeoJSON file.

Converting the route to a shapefile

The raster version of the least cost path route is useful for visualization, but it isn't much good for analysis because it is embedded in the raster, and it is, therefore, difficult to relate to other datasets as we have done so many other times in this book. Our next goal will be to use the path data that we saved when creating the route to create a shapefile since the saved data is in the proper order. The following code will convert our raster path to a shapefile that is easier to use in a GIS for analysis:

1. First, we'll import the modules that we need, which aren't many. We'll use the `pickle` module to restore the path `data` object. Then, we'll use the `linecache` module to read the geospatial header information from the path raster in order to map the path rows and columns to the earth coordinates. Finally, we'll use the `shapefile` module to export the shapefile:

   ```
   import pickle
   from linecache import getline
   import shapefile
   ```

2. Next, we'll create a function to convert rows and columns to x and y coordinates. The function accepts the metadata header information from the path raster file, as well as the column and row number:

   ```
   def pix2coord(gt,x,y):
     geotransform = gt
     ox = gt[2]
     oy = gt[3]
     pw = gt[4]
     ph = gt[4]
     cx = ox + pw * x + (pw/2)
     cy = oy + pw * y + (ph/2)
     return cx, cy
   ```

3. Now, we'll restore the `path` object from the pickled object:

   ```
   with open("path.p", "rb") as pathFile:
     path = pickle.load(pathFile)
   ```

4. Then, we'll parse the metadata information from the path raster file:

```
hdr = [getline("path.asc", i) for i in range(1, 7)]
gt = [float(ln.split(" ")[-1].strip()) for ln in hdr]
```

5. Next, we need a list object to hold the converted coordinates:

```
coords = []
```

6. Now, we convert each raster location from the least cost path object into a geospatial coordinate and store it in the list that we created:

```
for y,x in path:
  coords.append(pix2coord(gt,x,y))
```

7. Finally, with just a few lines, we write out a line shapefile:

```
with shapefile.Writer("path", shapeType=shapefile.POLYLINE) as w:
  w.field("NAME")
  w.record("LeastCostPath")
  w.line([coords])
```

Good work! You have created a program that can automatically navigate through obstacles, based on a set of rules, and exported it to a file that you can display and analyze in a GIS! We only used three rules, but you can add additional restrictions on how the program picks a path by adding other datasets, such as weather or water bodies, or anything else you can imagine.

Now that we understand blazing a path across an arbitrary surface, we'll look at routing through a network.

Routing along streets

Routing along streets uses a connected network of lines, which is called a graph. The lines in the graph can have impedance values, which discourage a routing algorithm from including them in a route. Examples of impedance values often include traffic volume, speed limit, or even distance. A key requirement for a routing graph is that all of the lines, known as edges, must be connected. Road datasets that are created for mapping will often have lines whose nodes do not intersect.

In this example, we'll calculate the shortest route through a graph by distance. We'll use a start and end point, which are not nodes in the graph, meaning we'll have to first find the graph nodes that are the closest to our start and destination points.

To calculate the shortest route, we'll use a powerful pure Python graph library called NetworkX. NetworkX is a general network graphing library that can create, manipulate, and analyze complex networks, including geospatial networks. If `pip` does not install NetworkX on your system, then you can find instructions for downloading and installing NetworkX for different operating systems at `http://networkx.readthedocs.org/en/stable/`.

You can download the road network and the start and end points, which are located along the U.S. Gulf Coast, as a ZIP file from `http://git.io/vcXFQ`. Then, you can follow these steps:

1. First, we'll need to import the libraries we're going to use. In addition to NetworkX, we'll use the PyShp library in order to read and write shapefiles:

```
import networkx as nx
import math
from itertools import tee
import shapefile
import os
```

2. Next, we'll define the current directory as our output directory for the route shapefile that we'll create:

```
savedir = "."
```

3. Now, we'll need a function that can calculate the distance between points in order to populate the impedance values of our graph and to find the nodes closest to our start and destination points for the route:

```
def haversine(n0, n1):
  x1, y1 = n0
  x2, y2 = n1
  x_dist = math.radians(x1 - x2)
  y_dist = math.radians(y1 - y2)
  y1_rad = math.radians(y1)
  y2_rad = math.radians(y2)
  a = math.sin(y_dist/2)**2 + math.sin(x_dist/2)**2 \
  * math.cos(y1_rad) * math.cos(y2_rad)
  c = 2 * math.asin(math.sqrt(a))
  distance = c * 6371
  return distance
```

4. Then, we'll create another function, which returns pairs of points from a list, to give us the line segments that we'll use to build our graph edges:

```
def pairwise(iterable):
 """Return an iterable in tuples of two
 s -> (s0,s1), (s1,s2), (s2, s3), ..."""
 a, b = tee(iterable)
 next(b, None)
 return zip(a, b)
```

5. Now, we'll define our road network shapefile. This road network is a subset of a U.S. interstate highway files shapefile from the **United States Geological Survey (USGS)**, which has been edited to ensure all the roads are connected:

```
shp = "road_network.shp"
```

6. Next, we'll create a graph with NetworkX and add the shapefile segments as graph edges:

```
G = nx.DiGraph()
r = shapefile.Reader(shp)
for s in r.shapes():
 for p1, p2 in pairwise(s.points):
 G.add_edge(tuple(p1), tuple(p2))
```

7. Then, we can extract the connected components as a subgraph. However, in this case, we've ensured that the entire graph is connected:

```
sg = list(nx.connected_component_subgraphs(G.to_undirected()))[0]
```

8. Next, we can read in the `start` and `end` points that we want to navigate:

```
r = shapefile.Reader("start_end")
start = r.shape(0).points[0]
end = r.shape(1).points[0]
```

9. Now, we loop through the graph, and assign distance values to each edge, using our haversine formula:

```
for n0, n1 in sg.edges_iter():
  dist = haversine(n0, n1)
  sg.edge[n0][n1]["dist"] = dist
```

10. Next, we must find the nodes in the graph that are the closest to our start and end points, in order to begin and end our route by looping through all of the nodes, and measuring the distance to our end points until we find the shortest distance:

```
nn_start = None
nn_end = None
start_delta = float("inf")
end_delta = float("inf")
for n in sg.nodes():
  s_dist = haversine(start, n)
  e_dist = haversine(end, n)
  if s_dist < start_delta:
  nn_start = n
  start_delta = s_dist
  if e_dist < end_delta:
  nn_end = n
  end_delta = e_dist
```

11. Now, we are ready to calculate the shortest distance through our road network:

```
path = nx.shortest_path(sg, source=nn_start, target=nn_end,
weight="dist")
```

12. Finally, we'll add the results to the shapefile and save our route:

```
w = shapefile.Writer(shapefile.POLYLINE)
w.field("NAME", "C", 40)
w.line(parts=[[list(p) for p in path]])
w.record("route")
w.save(os.path.join(savedir, "route"))
```

The following screenshot shows the road network in light gray, the start and end points, and the route in black. You can see that the route cuts across the road network in order to reach the road that is the nearest to the end point in the shortest possible distance:

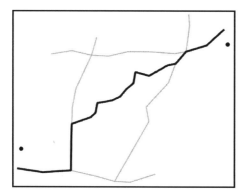

Now that we know how to create various types of routes, we can look at locating photos that you might take while traveling along a route.

Geolocating photos

Photos that are taken with GPS-enabled cameras, including smartphones, store location information in the header of the file in a format called **EXIF** tags. These tags are based largely on the same header tags that are used by the TIFF image standard. In this example, we'll use those tags to create a shapefile with point locations for the photos, and file paths to the photos, as attributes.

We'll use the PIL in this example because it has the ability to extract EXIF data. Most photos that are taken with smartphones are geotagged images; however, you can download the set used in this example from `https://git.io/vczR0`:

1. First, we'll import the libraries that we need, including PIL for the image metadata and PyShp for the shapefiles:

```
import glob
import os
try:
 import Image
 import ImageDraw
except ImportError:
 from PIL import Image
 from PIL.ExifTags import TAGS
import shapefile
```

2. Now, we'll need three functions. The first function extracts the EXIF data. The second function converts **degree, minutes, seconds (DMS)** coordinates to decimal degrees (EXIF data stores GPS data as DMS coordinates). The third function extracts the GPS data and performs the coordinate conversion:

```
def exif(img):
 # extract exif data.
 exif_data = {}
 try:
 i = Image.open(img)
 tags = i._getexif()
 for tag, value in tags.items():
 decoded = TAGS.get(tag, tag)
 exif_data[decoded] = value
 except:
 pass
 return exif_data

def dms2dd(d, m, s, i):
 # convert degrees, min, sec to decimal degrees
 sec = float((m * 60) + s)
 dec = float(sec / 3600)
 deg = float(d + dec)
 if i.upper() == 'W':
 deg = deg * -1
 elif i.upper() == 'S':
 deg = deg * -1
 return float(deg)

def gps(exif):
```

```
# get gps data from exif
lat = None
lon = None
if exif['GPSInfo']:
# Lat
coords = exif['GPSInfo']
i = coords[1]
d = coords[2][0][0]
m = coords[2][1][0]
s = coords[2][2][0]
lat = dms2dd(d, m, s, i)
# Lon
i = coords[3]
d = coords[4][0][0]
m = coords[4][1][0]
s = coords[4][2][0]
lon = dms2dd(d, m, s, i)
return lat, lon
```

3. Next, we will loop through the photos, extract the coordinates, and store the coordinates and filename in a dictionary:

```
photos = {}
photo_dir = "./photos"
files = glob.glob(os.path.join(photo_dir, "*.jpg"))
for f in files:
 e = exif(f)
 lat, lon = gps(e)
 photos[f] = [lon, lat]
```

4. Now, we will save the photo information as a shapefile:

```
with shapefile.Writer("photos", shapefile.POINT) as w:
    w.field("NAME", "C", 80)
    for f, coords in photos.items():
        w.point(*coords)
        w.record(f)
```

The filenames of the photos in the shapefile are now attributes of the point locations where the photos were taken. GIS programs including QGIS and ArcGIS have the tools to turn those attributes into links when you click on the photo path or the point. The following screenshot from QGIS shows that one of the photos opens after clicking on the associated point using the **Run Feature Action** tool:

To view the result, please use the following instructions:

1. Download QGIS from `https://qgis.org` and follow the installation instructions.
2. Open QGIS and drag the `photos.shp` file onto the blank map.
3. In the **Layer** panel on the left, right-click the layer named **Photos** and select **Properties**.
4. On the **Actions** tab, click the green plus sign to open the new actions dialog.
5. In the **Type** drop-down menu, select **Open**.
6. In the **Description** field, enter **Open Image**.
7. Click the **Insert** button in the lower-right corner.
8. Click the **OK** button, and then close the properties dialog.

9. Click on the small black arrow to the right of the **Run Feature Action** tool, which is a gear icon with a green center and a white arrow in it.

10. In the menu that pops up, choose **Open Image**.

11. Now, click on one of the points on the map to see the geotagged image popup.

Now, let's move from an image taken on the Earth, to images taken of the Earth itself, by working with satellite images.

Calculating satellite image cloud cover

Satellite images give us a powerful bird's-eye view of the Earth. They are useful for a variety of purposes, which we saw in Chapter 6, *Python and Remote Sensing*. However, they have one flaw—clouds. As a satellite passes around the Earth and collects imagery, it inevitably images clouds. And in addition to obstructing our view of the Earth, the cloud data can adversely affect remote sensing algorithms by wasting CPU cycles on useless cloud data, or skew the results by introducing unwanted data values.

The solution is to create a cloud mask. A cloud mask is a raster that isolates the cloud data in a separate raster. You can then use that raster as a reference when processing the image in order to avoid cloud data, or you can even use it to remove the clouds from the original image.

In this section, we'll create a cloud mask for a Landsat image using the rasterio module and the rio-l8qa plugin. The cloud mask will be created as a separate image that just contains clouds:

1. First, we need to download some sample Landsat 8 satellite image data as a ZIP file from http://bit.ly/landsat8data.

2. Click the download icon in the top right to download the data as a ZIP file, and unzip it to a directory named l8.

3. Next, make sure you have the raster libraries that we need by running pip:

   ```
   pip install rasterio
   pip install rio-l8qa
   ```

4. Now, we'll create the cloud mask by first importing the libraries that we need:

   ```
   import glob
   import os
   import rasterio
   from l8qa.qa import write_cloud_mask
   ```

5. Next, we need to provide a reference to our satellite image directory:

```
# Directory containing landsat data
landsat_dir = "18"
```

6. Now, we need to locate the quality-assurance metadata for the satellite data, which gives us the information that we need to generate the cloud mask:

```
src_qa = glob.glob(os.path.join(landsat_dir, '*QA*'))[0]
```

7. Finally, we use the quality-assurance file to create a cloud mask TIFF file:

```
with rasterio.open(src_qa) as qa_raster:
 profile = qa_raster.profile
 profile.update(nodata=0)
 write_cloud_mask(qa_raster.read(1), profile, 'cloudmask.tif')
```

The following image is just the band 7 (short-wave infrared) image from the Landsat 8 dataset:

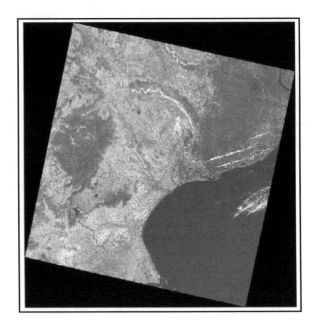

The next image is the cloud mask image containing only the location of clouds and shadows:

And finally, here's the mask over the image, showing the clouds as black:

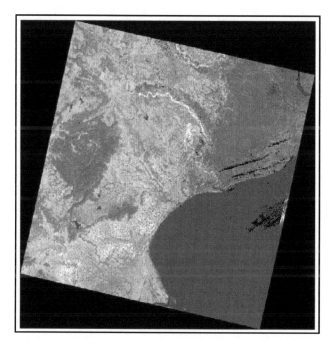

This example brushes the surface of what you can do with image masking. Another `rasterio` module, `rio-cloudmask`, allows you to calculate the cloud mask from scratch without using the quality-assurance data. But it requires some additional pre-processing steps. You can learn more about that here: `https://github.com/mapbox/rio-cloudmask`.

Summary

In this chapter, we learned how to create three real-world products, which are used every day in government, science, and industry. Apart from where this analysis is typically done with **black box** packages—costing thousands of dollars—we were able to use very minimal and free cross-platform Python tools. And in addition to the examples in this chapter, you now have some more reusable functions, algorithms, and processing frameworks for other advanced analyses, which will allow you to solve new problems that you come across in fields such as transportation, agriculture, and weather.

In the next chapter, we'll move into a relatively new area of geospatial analysis: real-time and near real-time data.

9
Real-Time Data

A common saying among geospatial analysts is: *A map is outdated as soon as it's created*. This saying reflects the fact that the Earth and everything on it are constantly changing. For most of the history of geospatial analysis and through most of this book, geospatial products are relatively static. Raw datasets are typically updated anywhere from a few months to a few years. The age of geospatial data in a map is referred to as **data currency**.

Data currency has traditionally not been the primary focus because of the time and expense needed to collect data. Web mapping, wireless cellular modems, and low-cost GPS antennas have changed that focus. It is now logistically feasible and even quite affordable to monitor a rapidly changing object or system and broadcast those changes to millions of people online. This change is revolutionizing geospatial technology and taking it in new directions. The most direct evidence of this revolution is web mapping mashups using systems such as Google Maps or OpenLayers and web-accessible data formats. Every day, more and more electronic devices are being brought online to broadcast their location and data for automation or remote control. Examples include thermostats, cameras, cars, and more. You can also use cheap, embedded computers such as the popular Raspberry Pi to turn almost anything into a connected **smart** device. This concept of connecting devices into a web of data and information is called **The Internet of Things (IoT)**.

In this chapter, we'll be checking out the following topics:

- Limitations of real-time data
- Using real-time data
- Tracking vehicles
- Storm chasing
- Reports from the field

By the end, you'll have learned to work with real-time geospatial data, and will be able to build a field reporting tool that can serve as a data transmission source for any type of data.

Technical requirements

This chapter requires the following things:

- Python 3.6 or higher
- RAM: Minimum 6 GB (Windows), 8 GB (macOS), recommended 8 GB
- Storage: Minimum 7200 RPM SATA with 20 GB of available space, and recommended SSD with 40 GB of available space
- Processor: Minimum Intel Core i3 2.5 GHz, and recommended Intel Core i5
- The MapQuest Developer API key, available here: `https://developer.mapquest.com/plan_purchase/steps/business_edition/business_edition_free/register`

Limitations of real-time data

The term **real-time data** typically means near-real-time. Some tracking devices capture real-time data and may update as often as several times a second. But the limitations of the infrastructure that broadcasts that data may constrain the output to every 10 seconds or longer. Weather radar is a perfect example. A **Doppler Weather Radar** (**DWR**) sweeps continuously but data is typically available online every five minutes. But given the contrast with traditional geospatial data updates, a refresh of a few minutes is real-time enough. Limitations can be summarized as follows:

- Network bandwidth limitations restricting data size
- Network latency limiting the data update frequency
- Availability of the data source due to restrictions such as battery life
- Lack of quality control due to data being instantly available to consumers
- Security vulnerabilities due to rapid ingestion of unverified data

Real-time data opens up additional opportunities for geospatial applications so we'll look at using it next.

Using real-time data

Web mashups often use real-time data. Web mashups are amazing and have changed the way many different industries operate. But they are typically limited in that they usually just display pre-processed data on a map and give developers access to a JavaScript API. But what if you want to process the data in some way? What if you want to filter, change, and then send it to another system? To use real-time data for geospatial analysis, you need to be able to access it as point data or a georeferenced raster.

 You can find out more about web map mashups here: `https://www.esri.com/arcgis-blog/products/product/uncategorized/digital-map-mashups/`.

As with examples in the previous chapters, the scripts are as simple as possible and designed to be read from start to finish without much mental looping. When functions are used they are listed first, followed by script variable declarations, and finally the main program execution.

Now let's see how to access a real-time and point-location data source using vehicles from the NextBus API.

Tracking vehicles

For our first real-time data source, we'll use the excellent **NextBus API**. NextBus (`http://www.nextbus.com/`) is a commercial service that tracks public transportation for municipalities including buses, trolleys, and trains. People riding these transit lines can then track the arrival time of the *next bus*.

What's even better is that, with the customer's permission, NextBus publishes tracking data through a **REpresentational State Transfer (REST) API.** Using URL API calls, developers can request information about a vehicle and receive an XML document about its location. This API is a straightforward way to begin using real-time data.

If you go to NextBus, you'll see a web interface as shown in the following screenshot, showing data for the city of Los Angeles, California metro system:

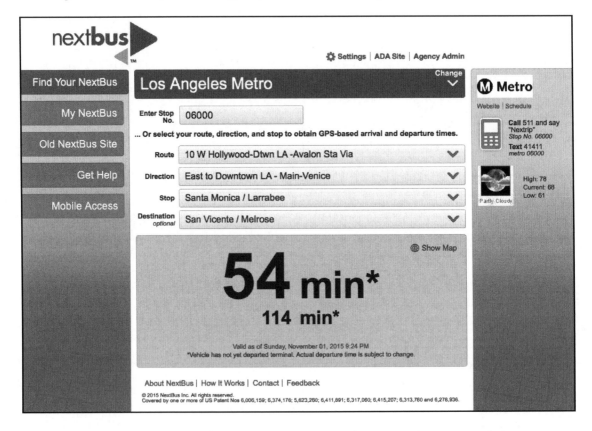

The system lets you select several parameters to learn the current location and time prediction for the next stop. On the right side of the screen, there is a link to a Google Maps mashup, showing transit tracking data for the particular route as shown in the following screenshot:

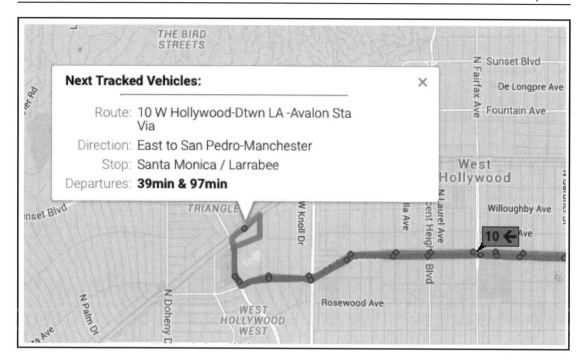

This is a very useful website but it does not give us control over how the data is displayed and used. Let's access the raw data directly using Python and the NextBus REST API to start working with real-time data.

For the examples in this chapter, we'll use the documented NextBus API found here: `http://www.nextbus.com/xmlFeedDocs/NextBusXMLFeed.pdf`.

To start with this example, we will need a list of buses required.

The NextBus agency list

NextBus customers are called **agencies**. In our examples, we are going to track buses on a route for Los Angeles, California. First, we need to get some information about the agency. The NextBus API consists of a web service named `publicXMLFeed`, in which you set a parameter named `command`. We'll call the `agencyList` command in a browser to get an XML document containing agency information using the following REST URL: `http://webservices.nextbus.com/service/publicXMLFeed?command=agencyList`.

When we go to that link in a browser, it returns an XML document containing the `<agency/>` tag. The tag for Los Angeles looks like the following:

```
<agency tag="lametro" title="Los Angeles Metro" regionTitle="California-
Southern"/>
```

Now that we have a list of buses, we need to get the routes they can travel.

The NextBus route list

The `tag` attribute is the ID for Thunder Bay, which we need for other NextBus API commands. The other attributes are human-readable metadata. The next piece of information we need is the details about the **route 2** bus route. To get this information, we'll use the agency ID and the `routeList` REST command to get another XML document by pasting the URL into our web browser.

 Note that the agency ID is set to the parameter in the REST URL: `http://webservices.nextbus.com/service/publicXMLFeed?command=routeList a=lametro`.

When we call this URL in a browser, we get the following XML document:

```
<?xml version="1.0" encoding="utf-8" ?>
<body copyright="All data copyright Los Angeles Metro 2015."><route tag="2"
title="2 Downtown LA - Pacific Palisades Via"/><route tag="4" title="4
Downtown LA - Santa Monica Via Santa"/>
<route tag="10" title="10 W Hollywood-Dtwn LA -Avalon Sta Via"/>
...
<route tag="901" title="901 Metro Orange Line"/>
<route tag="910" title="910 Metro Silver Line"/>
</body>
```

We have buses and routes. We're ready to start tracking their locations!

NextBus vehicle locations

So the mainline route ID stored in the `tag` attribute is simply 1, according to these results. Thus, now, we have all of the information we need to track buses along the **LA Metro route 2**.

There is only one more required parameter (called `t`) that represents milliseconds since the **1970 epoch date (January 1, 1970, at midnight UTC)**. The epoch date is simply a computer standard used by machines to track time. The easiest thing to do within the NextBus API is to specify `0` for this value, which returns data for the last 15 minutes.

There is an optional `direction` tag that allows you to specify a terminating bus stop in the event a route has multiple buses running on the route in opposite directions. But, if we don't specify that, the API will return the first one, which suits our needs. The REST URL to get the mainline route for LA Metro looks like the following: `http://webservices.nextbus.com/service/publicXMLFeed?command=vehicleLocationsa=lametror=2t=0`.

Calling this REST URL in a browser returns the following XML document:

```
<?xml version="1.0" encoding="utf-8" ?>
<body copyright="All data copyright Los Angeles Metro 2015.">
<vehicle id="7582" routeTag="2" dirTag="2_758_0" lat="34.097992"
lon="-118.350365" secsSinceReport="44" predictable="true" heading="90"
speedKmHr="0"/>
<vehicle id="7583" routeTag="2" dirTag="2_779_0" lat="34.098076"
lon="-118.301399" secsSinceReport="104" predictable="true" heading="90"
speedKmHr="37"/>
. . .
</body >
```

Each `vehicle` tag represents a location within the last 15 minutes. The `last` tag is the most recent location (even though XML is technically unordered).

These public transportation systems do not run all of the time. Many close down at 10:00 p.m. (22:00) local time. If you encounter an error in the script, use the NextBus website to locate a system that is running and change the agency and route variables to that system.

We can now write a Python script that returns the locations for a bus on a given route. If we don't specify the `direction` tag, NextBus returns the first one. In this example, we are going to poll the NextBus tracking API by calling the REST URL using the built-in Python `urllib` library demonstrated in previous chapters.

We'll parse the returned XML document using the simple built-in `minidom` module, also shown in *The minidom module* section, in `Chapter 4`, *Geospatial Python Toolbox*. This script simply outputs the latest latitude and longitude of the route 2 bus. You will see the agency and route variables near the top. To do this, we need to follow the following steps:

1. First, we import the libraries we need:

```
import urllib.request
import urllib.parse
import urllib.error
from xml.dom import minidom
```

2. Now we set up our variables for API mode and the customer and route we want to query:

```
# Nextbus API command mode
command = "vehicleLocations"

# Nextbus customer to query
agency = "lametro"

# Bus we want to query
route = "2"
```

3. We're going to set the time value to 0, which will grab the last 15 minutes of data:

```
# Time in milliseconds since the
# 1970 epoch time. All tracks
# after this time will be returned.
# 0 only returns data for the last
# 15 minutes
epoch = "0"
```

4. Now we need to build the query URL that we'll use to access the API:

```
# Build our query url
# webservices base url
url = "http://webservices.nextbus.com"

# web service path
url += "/service/publicXMLFeed?"

# service command/mode
url += "command=" + command

# agency
```

```
url += "&a=" + agency
url += "&r=" + route
url += "&t=" + epoch
```

5. Next, we can call the API using `urllib`:

```
# Access the REST URL
feed = urllib.request.urlopen(url)
if feed:
 # Parse the xml feed
 xml = minidom.parse(feed)
 # Get the vehicle tags
 vehicles = xml.getElementsByTagName("vehicle")
 # Get the most recent one. Normally there will
 # be only one.
```

6. Finally, we can access the results and print out the location of each bus:

```
if vehicles:
  bus = vehicles.pop()
  # Print the bus latitude and longitude
  att = bus.attributes
  print(att["lon"].value, ",", att["lat"].value)
else:
  print("No vehicles found.")
```

The output of this script is simply a latitude and longitude value that implies that we now have control of the API and understand it. The output should be a coordinate value for the latitude and longitude.

Now we are ready to use these location values to create our own map.

Mapping NextBus locations

The best source of freely available street mapping data is the **OpenStreetMap (OSM)** project: http://www.openstreetmap.org. OSM also has a publicly available REST API for creating static map images called **StaticMapLite**: http://staticmap.openstreetmap.de.

The **OSM StaticMapLite API** provides a GET API based on Google's static map API to create simple map images with a limited number of point markers and lines. A GET API, as opposed to a REST, API allows you to append name/value parameter pairs after a question mark on the URL. A REST API makes the parameters part of the URL path. We'll use the API to create our own NextBus API map on-demand with a red pushpin icon for the bus location.

In the next example, we have condensed the previous script down to a compact function named nextbus(). The nextbus() function accepts an agency, route, command, and epoch as arguments. The command defaults to vehicleLocations and the epoch defaults to 0 to get the last 15 minutes of data. In this script, we'll pass in the LA route-2 route information and use the default command that returns the most recent latitude/longitude of the bus.

We have a second function named nextmap() that creates a map with a purple marker on the current location of the bus each time it is called. The map is created by building a GET URL for the OSM StaticMapLite API, which centers on the location of the bus and uses a zoom level between *1-18* and the map size to determine the map extent.

You can access the API directly in a browser to see an example of what the nextmap() function does. You will need a free MapQuest Developer API key available by registering here: https://developer.mapquest.com/ plan_purchase/steps/business_edition/business_edition_free/ register. Once you have the key, insert it in the key parameter where it says YOUR_API_KEY_HERE. Then, you can test the following example URL: https://www.mapquestapi.com/staticmap/v4/getmap?size= 865,512&type=map&pois=mcenter,40.702147,-74.015794|&zoom= 14&center=40.714728,-73.998672&imagetype=JPEG&key=YOUR_AP I_KEY_HERE.

Static maps look similar to the following:

The nextmap() function accepts a NextBus agency ID, route ID, and string for the base image name for the map. The function calls the nextbus() function to get the latitude/longitude pair. The execution of this program loops through at timed intervals, creates a map on the first pass, and then overwrites the map on subsequent passes. The program also outputs a timestamp each time a map is saved. The requests variable specifies the number of passes and the freq variable represents the time in seconds between each loop. Let's check the following code to see how of this example works:

1. First, we import the libraries we need:

```
import urllib.request
import urllib.parse
import urllib.error
from xml.dom import minidom
import time
```

2. Next, we create a function that can get the latest location of a bus on a given route:

```
def nextbus(a, r, c="vehicleLocations", e=0):
 """Returns the most recent latitude and
 longitude of the selected bus line using
 the NextBus API (nbapi)
 Arguments: a=agency, r=route, c=command,
 e=epoch timestamp for start date of track,
 0 = the last 15 minutes"""
 nbapi = "http://webservices.nextbus.com"
 nbapi += "/service/publicXMLFeed?"
 nbapi += "command={}&a={}&r={}&t={}".format(c, a, r, e)
 xml = minidom.parse(urllib.request.urlopen(nbapi))
 # If more than one vehicle, just get the first
 bus = xml.getElementsByTagName("vehicle")[0]
 if bus:
 at = bus.attributes
 return(at["lat"].value, at["lon"].value)
 else:
 return (False, False)
```

3. Now we have a function to plot a bus location on a map image:

```
def nextmap(a, r, mapimg):
 """Plots a nextbus location on a map image
 and saves it to disk using the MapQuest OpenStreetMap Static Map
 API (osmapi)"""
 # Fetch the latest bus location
 lat, lon = nextbus(a, r)
 if not lat:
   return False
 # Base url + service path
```

4. Within that function, we set up the API parameters in the URL:

```
 osmapi = "https://www.mapquestapi.com/staticmap/v4/getmap?
 type=map&"
 # Use a red, pushpin marker to pin point the bus
 osmapi += "mcenter={},{}|&".format(lat, lon)
 # Set the zoom level (between 1-18, higher=lower scale)
 osmapi += "zoom=18&"
 # Center the map around the bus location
 osmapi += "center={},{}&".format(lat, lon)
 # Set the map image size
 osmapi += "&size=1500,1000"
 # Add our API Key
 osmapi += "&key=YOUR_API_KEY_HERE"
```

5. Now we can create the image by calling the URL and save it:

```
# Create a PNG image
osmapi += "imagetype=png&"
img = urllib.request.urlopen(osmapi)

# Save the map image
with open("{}.png".format(mapimg), "wb") as f:
    f.write(img.read())
return True
```

6. Now in our main program, we can set up variables about the buses we want to track:

```
# Nextbus API agency and bus line variables
agency = "lametro"
route = "2"
# Name of map image to save as PNG
nextimg = "nextmap"
```

7. Then, we can specify the number and frequency of tracking points we want:

```
# Number of updates we want to make
requests = 1
# How often we want to update (seconds)
freq = 5
```

8. Finally, we can begin tracking and updating our map image:

```
# Map the bus location every few seconds
for i in range(requests):
 success = nextmap(agency, route, nextimg)
 if not success:
   print("No data available.")
   continue
 print("Saved map {} at {}".format(i, time.asctime()))
 time.sleep(freq)
```

9. While the script runs, you'll see an output similar to the following, showing at what time the script saved each map:

```
Saved map 0 at Sun Nov 1 22:35:17 2015
Saved map 1 at Sun Nov 1 22:35:24 2015
Saved map 2 at Sun Nov 1 22:35:32 2015
```

This script saves a map image similar to the following, depending on where the bus was when you ran it:

This map is an excellent example of using an API to create a custom mapping product. But it is a very basic tracking application. To begin to develop it into a more interesting geospatial product, we need to combine it with some other real-time data source that gives us more situational awareness.

Now that we can track buses, let's add some additional information to the map that would be useful to know for passengers taking a bus. Let's add some weather data.

Storm chasing

So far, we have created a simpler version of what the NextBus website already does. But we have done it in a way that ultimately gives us complete control over the output. Now we want to use this control to go beyond what the NextBus Google Maps mashup does. We'll add another real-time data source that is very important to both travelers and bus-line operators: the weather.

Iowa State University's Mesonet program provides free and polished weather data for applications. We use this data to create a real-time weather map for our bus location map. We can use the **Open Geospatial Consortium (OGC) Web Map Service (WMS)** standard to request a single image over our area of interest. A WMS is an OGC standard for serving georeferenced map images through the web; they are generated by a map server through an HTTP request.

The Mesonet system provides an excellent web mapping service that returns a subsetted image from a global precipitation mosaic based on a properly-formatted WMS request. An example of such a request is the following query: `http://mesonet.agron.iastate.edu/ cgi-bin/wms/nexrad/n0r.cgi?SERVICE=WMSVERSION=1.1.1REQUEST=GetMapLAYERS=nexrad-n0rSTYLES=SRS=EPSG:900913BBOX=-15269659.42,2002143.61,-6103682.81,7618920.15 WIDTH=600HEIGHT=600FORMAT=image/png`.

Because the examples in this chapter rely on real-time data, the specific requests listed may produce blank weather images if there is no activity in the area of interest. You can visit this link (http://radar.weather.gov/ridge/Conus/index.php) to find an area where a storm is occurring. This page contains a KML link for Google Earth or QGIS. These WMS images are transparent PNG images similar to the following sample:

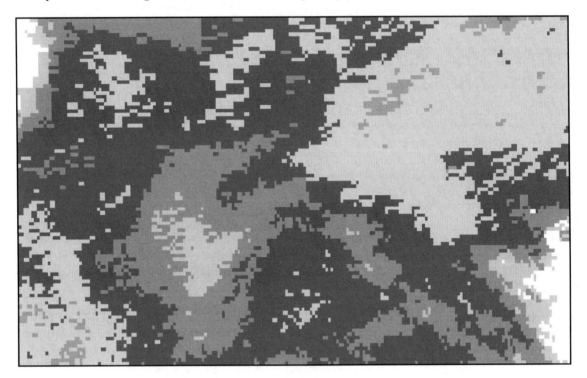

The OSM site, on the other hand, no longer provides its street maps via WMS—only as tiles. They do, however, allow other organizations to download tiles or raw data to extend the free service. The US **National Oceanic and Atmospheric Administration** (**NOAA**) has done just that and provided a WMS interface to their OSM data, allowing requests to retrieve the single basemap image we need for our bus route:

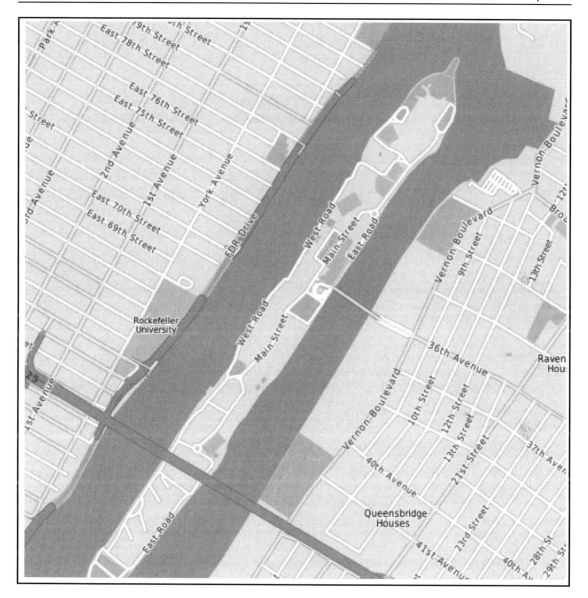

We now have data sources from which to get the basemap and weather data. We want to combine these images and plot the current location of the bus. Instead of a simple dot, we'll get a little more sophisticated and add the following bus icon this time:

You will need to download this icon, `busicon.png`, to your working directory from here: `https://github.com/GeospatialPython/Learn/blob/master/busicon.png?raw=true`.

Now we'll combine our previous scripts and our new data sources to create a real-time weather bus map. Because we are going to blend the street map and weather map, we'll need the **Python Imaging Library (PIL)** used in previous chapters. We'll replace our `nextmap()` function from the previous example with a simple `wms()` function that can grab a map image by a bounding box from any WMS service. We'll also add a function that converts decimal degrees into meters, named `ll2m()`.

The script gets the bus location, converts the location to meters, creates a 2 mile (3.2 km) rectangle around the location, and then downloads a street and weather map. The map images are then blended together using PIL. PIL then shrinks the bus icon image to 30 x 30 pixels and pastes it in the center of the map, which is the bus location. Let's look at how the following code works:

1. First, we'll import the libraries we need:

```
import sys
import urllib.request
import urllib.parse
import urllib.error
from xml.dom import minidom
import math
try:
  import Image
```

```
except:
   from PIL import Image
```

2. Now we'll reuse our `nextbus` function from the previous example to get the bus tracking data:

```python
def nextbus(a, r, c="vehicleLocations", e=0):
 """Returns the most recent latitude and
 longitude of the selected bus line using
 the NextBus API (nbapi)"""
 nbapi = "http://webservices.nextbus.com"
 nbapi += "/service/publicXMLFeed?"
 nbapi += "command=%s&a=%s&r=%s&t=%s" % (c, a, r, e)
 xml = minidom.parse(urllib.request.urlopen(nbapi))
 # If more than one vehicle, just get the first
 bus = xml.getElementsByTagName("vehicle")[0]
 if bus:
 at = bus.attributes
 return(at["lat"].value, at["lon"].value)
 else:
 return (False, False)
```

3. We also need a function to convert latitude and longitude into meters:

```python
def ll2m(lon, lat):
 """Lat/lon to meters"""
 x = lon * 20037508.34 / 180.0
 y = math.log(math.tan((90.0 + lat) *
 math.pi / 360.0)) / (math.pi / 180.0)
 y = y * 20037508.34 / 180
 return (x, y)
```

4. Now we need a function to retrieve WMS map images, which we'll use for our weather image:

```python
def wms(minx, miny, maxx, maxy, service, lyr, epsg, style, img, w,
        h):
    """Retrieve a wms map image from
    the specified service and saves it as a JPEG."""
    wms = service
    wms += "?SERVICE=WMS&VERSION=1.1.1&REQUEST=GetMap&"
    wms += "LAYERS={}".format(lyr)
    wms += "&STYLES={}&".format(style)
    wms += "SRS=EPSG:{}&".format(epsg)
    wms += "BBOX={},{},{},{}&".format(minx, miny, maxx, maxy)
    wms += "WIDTH={}&".format(w)
    wms += "HEIGHT={}&".format(h)
    wms += "FORMAT=image/jpeg"
```

```
    wmsmap = urllib.request.urlopen(wms)
    with open(img + ".jpg", "wb") as f:
        f.write(wmsmap.read())
```

5. Now we can set up all of the variables in our main program to use our functions:

```
# Nextbus agency and route ids
agency = "roosevelt"
route = "shuttle"
# OpenStreetMap WMS service
basemap = "http://ows.mundialis.de/services/service"
# Name of the WMS street layer
streets = "TOPO-OSM-WMS"
# Name of the basemap image to save
mapimg = "basemap"
# OpenWeatherMap.org WMS Service
weather =
"https://mesonet.agron.iastate.edu/cgi-bin/wms/nexrad/n0q.cgi?"
# If the sky is clear over New York,
# use the following url which contains
# a notional precipitation sample:
# weather = "http://git.io/vl4r1"
# WMS weather layer
weather_layer = "nexrad-n0q-900913"
# Name of the weather image to save
skyimg = "weather"
# Name of the finished map to save
final = "next-weather"
# Transparency level for weather layer
# when we blend it with the basemap.
# 0 = invisible, 1 = no transparency
opacity = .5
# Pixel width and height of the
# output map images
w = 600
h = 600
# Pixel width/height of the the
# bus marker icon
icon = 30
```

6. Now we're ready to get our bus location:

```
# Get the bus location
lat, lon = nextbus(agency, route)
if not lat:
 print("No bus data available.")
 print("Please try again later")
 sys.exit()
# Convert strings to floats
lat = float(lat)
lon = float(lon)
# Convert the degrees to Web Mercator
# to match the NOAA OSM WMS map
x, y = ll2m(lon, lat)
# Create a bounding box 1600 meters
# in each direction around the bus
minx = x - 1600
maxx = x + 1600
miny = y - 1600
maxy = y + 1600
```

7. Then, we can download our street map:

```
# Download the street map
wms(minx, miny, maxx, maxy, basemap, streets, mapimg, w, h)
```

8. Then, we can download the weather map:

```
# Download the weather map
wms(minx, miny, maxx, maxy, weather, weather_layer, skyimg, w, h)
```

9. Now we can overlay the weather data on the bus map:

```
# Open the basemap image in PIL
im1 = Image.open("basemap.png").convert('RGBA')
# Open the weather image in PIL
im2 = Image.open("weather.png").convert('RGBA')
# Convert the weather image mode
# to "RGB" from an indexed PNG
# so it matches the basemap image
im2 = im2.convert(im1.mode)
# Create a blended image combining
# the basemap with the weather map
im3 = Image.blend(im1, im2, opacity)
```

10. Next, we need to add the bus icon to our combined map to show the bus's location:

```
# Open up the bus icon image to
# use as a location marker.
# http://git.io/vlgHl
im4 = Image.open("busicon.png")
# Shrink the icon to the desired
# size
im4.thumbnail((icon, icon))
# Use the blended map image
# and icon sizes to place
# the icon in the center of
# the image since the map
# is centered on the bus
# location.
w, h = im3.size
w2, h2 = im4.size
# Paste the icon in the center of the image
center_width = int((w/2)-(w2/2))
center_height = int((h/2)-(h2/2))
im3.paste(im4, (center_width, center_height), im4)
```

11. Finally, we can save the finished map:

```
# Save the finished map
im3.save(final + ".png")
```

This script will produce a map similar to the following:

The map shows us that the bus is experiencing moderate precipitation at its current location. The color ramp, as shown in the Mesonet website screenshot earlier, ranges from light blue for light precipitation, then green, yellow, orange, to red as the rain gets heavier (or light gray to darker gray in black and white). So, at the time this map was created, the bus-line operator could use this image to tell their drivers to go a little slower, and passengers will know they may want to get an umbrella before heading to the bus stop.

Because we wanted to learn the NextBus API at a low level, we used the API directly using built-in Python modules. But several third-party Python modules exist for the API including one on PyPI, simply called `nextbus`, which allows you to work with higher-level objects for all of the NextBus commands and provides more robust error handling not included in the simple examples in this chapter.

Now that we've learned how to check the weather, let's combine discrete real-time data sources into more meaningful products using Python, HTML, and JavaScript.

Reports from the field

In our final example in this chapter, we'll get off of the bus and out into the field. Modern smartphones, tablets, and laptops allow us to update a GIS and view those updates from everywhere. We'll use HTML, GeoJSON, the Leaflet JavaScript library, and a pure-Python library named Folium to create a client-server application that allows us to post geospatial information to a server and then create an interactive web map to view those data updates.

First, we need a web form that shows your current location and updates the server when you submit the form with comments about your location. You can find the form here: `http://geospatialpython.github.io/Learn/fieldwork.html`.

The following screenshot shows the form:

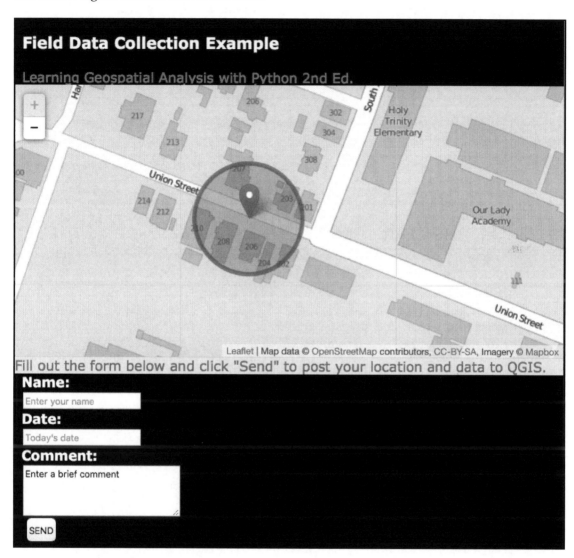

You can view the source of that form to see how it works. The mapping is done using the Leaflet library and posts GeoJSON to a unique URL on myjson.com. You can use this page on mobile devices, move it to any web server, or even use it on your local hard drive.

The form posts to the following URL publicly on `myjson.com`: `https://api.myjson.com/bins/467pm`. You can visit that URL in a browser to see the raw GeoJSON.

Next, you need to install the Folium library from PyPI. Folium provides a simple Python API for creating Leaflet web maps. You can find more information about Folium here: `https://github.com/python-visualization/folium`.

Folium makes producing a Leaflet map extremely simple. This script is just a few lines and will output a web page named `map.html`. We pass the GeoJSON URL to the `map` object, which will plot the locations on the map:

```
import folium
m = folium.Map()
m.geo_json(geo_path="https://api.myjson.com/bins/467pm")
m.create_map(path="map.html")
```

The resulting interactive map will display points as markers. When you click on a marker, the information from the form is displayed. You can just open the HTML file in any browser.

Summary

Real-time data is an exciting way to do new types of geospatial analysis, only recently made possible by advances in several different technologies, including web mapping, GPS, and wireless communications. In this chapter, you learned how to access raw feeds for real-time location data, how to acquire a subset of real-time raster data, how to combine different types of real-time data into a custom map analysis product using only Python, and how to build client-server geospatial applications to update a GIS in real-time.

As with previous chapters, these examples contain building blocks that will let you build new types of application using Python that go far beyond the typical popular and ubiquitous JavaScript-based mashups.

In the next chapter, we will combine everything we've learned so far into a complete geospatial application that applies algorithms and concepts in a realistic scenario.

10
Putting It All Together

Throughout the book, we have touched all the important aspects of geospatial analysis and we've used a variety of different techniques in Python to analyze different types of geospatial data. In this final chapter, we will draw on nearly all of the topics we have covered to produce a real-world product that has become very popular: a GPS route analysis report.

These reports are common to dozens of mobile app services, GPS watches, in-car navigation systems, and other GPS-based tools. A GPS typically records location, time, and elevation. From these values, we can derive a vast amount of ancillary information about what happened along the route on which that data was recorded. Fitness apps including RunKeeper, MapMyRun, Strava, and Nike Plus all use similar reports to present GPS-tracked exercise data from running, hiking, biking, and walking.

We will create one of these reports using Python. This program is nearly 500 lines of code, our longest yet, so we will step through it in pieces. We will combine the following techniques:

- Understanding a typical GPS report
- Building a GPS reporting tool

As we step through this program, all of the techniques used will be familiar, but we will be using them in new ways.

Technical requirements

We'll be needing the following things for this chapter:

- Python 3.6 or higher
- RAM: Minimum – 6 GB (Windows), 8 GB (macOS); recommended 8 GB
- Storage: Minimum 7200 RPM SATA with 20 GB of available space, recommended SSD with 40 GB of available space

- Processor: Minimum Intel Core i3 2.5 GHz, recommended Intel Core i5
- PIL: The Python Imaging Library
- NumPy: A multidimensional and array-processing library
- `pygooglechart`: A Python wrapper for the excellent Google Chart API
- FPDF: A simple and pure-Python PDF writer

Understanding a typical GPS report

A typical GPS report has common elements including a route map, elevation profile, and speed profile. The following screenshot is a report from a typical route logged through RunKeeper (`https://runkeeper.com/index`):

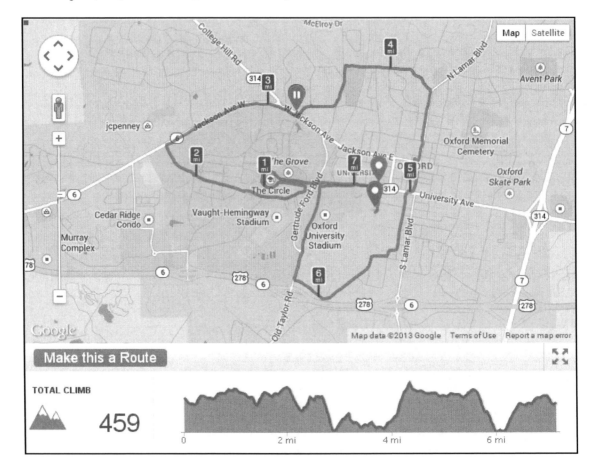

Our report will be similar, but we'll add a twist. We'll include the route map and elevation profile like this service, but we'll also add the weather conditions that occurred on that route when it was recorded and a geolocated photo taken on route.

Now that we know what a GPS report is, let's learn how to build it.

Building a GPS reporting tool

The name of our program is `GPX-Reporter.py`. If you remember the *Tag and markup-based formats* section in `Chapter 2`, *Learning Geospatial Data*, the **GPX** format is the most common way to store GPS route information. Nearly every program and device relying on GPS data can convert to and from GPX.

For this example, you can download a sample GPX file from: `http://git.io/vl7qi`. Also, you will need to install a few Python libraries from PyPI.

You should simply use `easy_install` or `pip` to install these tools. We will also be using a module called `SRTM.py`. This module is a utility for working with near-global elevation data collected during the 11-day **Shuttle Radar Topography Mission (SRTM)** in 2000 by the space shuttle Endeavor. Install the SRTM module using `pip`:

```
pip install srtm.py
```

Alternatively, you can also download the zipped file, extract it, and copy the `srtm` folder to your Python `site-packages` directory or your working directory: `http://git.io/vl5Ls`.

You will also need to register for a free Dark Sky API. This free service provides unique tools. It is the only service that provides global, historical weather data for nearly any point location with up to 1,000 requests per day for free: `https://darksky.net/dev`.

Dark Sky will provide you with a text key that you insert into a variable called `api_key` in the GPX-Reporter program before running it. Finally, as per Dark Sky's terms of service, you'll need to download a logo image to be inserted into the report: `https://raw.githubusercontent.com/GeospatialPython/Learn/master/darksky.png`.

 You can review the Dark Sky *Terms of Service* here: `https://darksky.net/dev/docs/terms`.

Now, we're ready to work through the GPX-Reporter program. Like other scripts in this book, this program tries to minimize functions so you can mentally trace the program better and modify it with less effort. The following list contains the major steps in the program:

1. Setting up the Python `logging` module
2. Establishing our helper functions
3. Parsing the GPX data file
4. Calculating the route bounding box
5. Buffering the bounding box
6. Converting the box to meters
7. Downloading the basemap
8. Downloading the elevation data
9. Hillshading the elevation data
10. Increasing the hillshade contrast
11. Blending the hillshade and basemap
12. Drawing the GPX track on a separate image
13. Blending the track image and basemap
14. Drawing the start and finish points
15. Saving the map image
16. Calculating the route mile markers
17. Building the elevation profile chart
18. Getting the weather data for the route time period
19. Generating the PDF report

The next subsection takes you through the first step.

Initial setup

The beginning of the program is `import` statements followed by the Python `logging` module. The `logging` module provides a more robust way to track and log program status than simple `print` statements. In this part of the program, we configure it as shown in the following steps:

1. We first need to install all the libraries we need, as shown in the following code:

```
from xml.dom import minidom
import json
```

```
import urllib.request
import urllib.parse
import urllib.error
import math
import time
import logging
import numpy as np
import srtm # Python 3 version: http://git.io/vl5Ls
import sys
from pygooglechart import SimpleLineChart
from pygooglechart import Axis
import fpdf
import glob
import os
try:
 import Image
 import ImageFilter
 import ImageEnhance
 import ImageDraw
except:
 from PIL import Image
 from PIL import ImageFilter
 from PIL import ImageEnhance
 from PIL import ImageDraw
 from PIL.ExifTags import TAGS
```

2. Now we can configure the Python `logging` module to tell us what's going on throughout the process, as shown here:

```
# Python logging module.
# Provides a more advanced way
# to track and log program progress.
# Logging level - everything at or below
# this level will output. INFO is below.
level = logging.DEBUG
# The formatter formats the log message.
# In this case we print the local time, logger name, and message
formatter = logging.Formatter("%(asctime)s - %(name)s -
%(message)s")
# Establish a logging object and name it
log = logging.getLogger("GPX-Reporter")
# Configure our logger
log.setLevel(level)
# Print to the command line
console = logging.StreamHandler()
console.setLevel(level)
console.setFormatter(formatter)
log.addHandler(console)
```

This logger prints to the console, but with a few simple modifications you can have it print to a file, or even a database, just by altering the configuration in this section. This module is built into Python and is documented here: `https://docs.python.org/3/howto/logging.html`.

Next, we have several utility functions that are used several times throughout the program.

Working with utility functions

All of the following functions, except the functions related to time, have been used in previous chapters in some form. Let's see how to use utility functions in our example:

1. First, the `l12m()` function converts latitude and longitude to meters:

```python
def l12m(lat, lon):
 """Lat/lon to meters"""
 x = lon * 20037508.34 / 180.0
 y = math.log(math.tan((90.0 + lat) *
 math.pi / 360.0)) / (math.pi / 180.0)
 y = y * 20037508.34 / 180
 return (x, y)
```

2. The `world2pixel()` function converts geospatial coordinates to pixel coordinates on our output map image:

```python
def world2pixel(x, y, w, h, bbox):
 """Converts world coordinates
 to image pixel coordinates"""
 # Bounding box of the map
 minx, miny, maxx, maxy = bbox
 # world x distance
 xdist = maxx - minx
 # world y distance
 ydist = maxy - miny
 # scaling factors for x, y
 xratio = w/xdist
 yratio = h/ydist
 # Calculate x, y pixel coordinate
 px = w - ((maxx - x) * xratio)
 py = (maxy-y) * yratio
 return int(px), int(py)
```

3. Then, we have `get_utc_epoch()` and `get_local_time()` to convert the UTC time stored in the GPX file to local time along the route:

```
def get_utc_epoch(timestr):
    """Converts a GPX timestamp to Unix epoch seconds
    in Greenwich Mean Time to make time math easier"""
    # Get time object from ISO time string
    utctime = time.strptime(timestr, '%Y-%m-%dT%H:%M:%S.000Z')
    # Convert to seconds since epoch
    secs = int(time.mktime(utctime))
    return secs
```

4. Now we have a haversine distance function and our simple `wms` function to retrieve map images:

```
def haversine(x1, y1, x2, y2):
    """Haversine distance formula"""
    x_dist = math.radians(x1 - x2)
    y_dist = math.radians(y1 - y2)
    y1_rad = math.radians(y1)
    y2_rad = math.radians(y2)
    a = math.sin(y_dist/2)**2 + math.sin(x_dist/2)**2 \
    * math.cos(y1_rad) * math.cos(y2_rad)
    c = 2 * math.asin(math.sqrt(a))
    # Distance in miles. Just use c * 6371
    # for kilometers
    distance = c * (6371/1.609344) # Miles
    return distance
```

5. The `wms()` function retrieves map images with the following code:

```
def wms(minx, miny, maxx, maxy, service, lyr, epsg, style, img, w,
h):
    """Retrieve a wms map image from
    the specified service and saves it as a JPEG."""
    wms = service
    wms += "?SERVICE=WMS&VERSION=1.1.1&REQUEST=GetMap&"
    wms += "LAYERS={}".format(lyr)
    wms += "&STYLES={}&".format(style)
    wms += "SRS=EPSG:{}&".format(epsg)
    wms += "BBOX={},{},{},{}&".format(minx, miny, maxx, maxy)
    wms += "WIDTH={}&".format(w)
    wms += "HEIGHT={}&".format(h)
    wms += "FORMAT=image/jpeg"
    wmsmap = urllib.request.urlopen(wms)
    with open(img + ".jpg", "wb") as f:
        f.write(wmsmap.read())
```

6. Next, we have an `exif()` function to extract the metadata from the photo:

```
def exif(img):
 """Return EXIF metatdata from image"""
 exif_data = {}
 try:
 i = Image.open(img)
 tags = i._getexif()
 for tag, value in tags.items():
 decoded = TAGS.get(tag, tag)
 exif_data[decoded] = value
 except:
 pass
 return exif_data
```

7. Then we have a `dms2dd()` function to convert degrees/minutes/seconds coordinates to decimal degrees because that's how the photo coordinates are stored:

```
def dms2dd(d, m, s, i):
 """Convert degrees/minutes/seconds to
 decimal degrees"""
 s *= .01
 sec = float((m * 60.0) + s)
 dec = float(sec / 3600.0)
 deg = float(d + dec)
 if i.upper() == 'W':
 deg = deg * -1.0
 elif i.upper() == 'S':
 deg = deg * -1.0
 return float(deg)
```

8. And finally, we have a `gps()` function to extract the coordinates from the photo metadata:

```
def gps(exif):
 """Extract GPS info from EXIF metadat"""
 lat = None
 lon = None
 if exif['GPSInfo']:
 # Lat
 coords = exif['GPSInfo']
 i = coords[1]
 d = coords[2][0][0]
 m = coords[2][1][0]
 s = coords[2][2][0]
 lat = dms2dd(d, m ,s, i)
 # Lon
```

```
i = coords[3]
d = coords[4][0][0]
m = coords[4][1][0]
s = coords[4][2][0]
lon = dms2dd(d, m ,s, i)
return lat, lon
```

9. Next, we have our program variables. We will be accessing an **OpenStreetMap WMS** service provided for free by a company named **Mundalis** as well as the SRTM data provided by NASA.

 We access the WMS services in this book using Python's `urllib` library for simplicity, but if you plan to use OGC web services frequently, you should use the Python package OWSLib available through PyPI: `https://pypi.python.org/pypi/OWSLib`.

Now let's perform the followings steps to set up the WMS web service:

1. We will output several intermediate products and images. These variables are used in those steps. The `route.gpx` file is defined in this section as the `gpx` variable. First, we set up some conversion constants for degrees to radians conversion and back with the following code:

```
# Needed for numpy conversions in hillshading
deg2rad = 3.141592653589793 / 180.0
rad2deg = 180.0 / 3.141592653589793
```

2. Next, we set up the name of our `.gpx` file as follows:

```
# Program Variables

# Name of the gpx file containing a route.
# https://git.io/fjwHW
gpx = "route.gpx"
```

3. Now, we begin setting up the WMS web service, which will retrieve the map:

```
# NOAA OpenStreetMap Basemap

# OSM WMS service
osm_WMS = "http://ows.mundialis.de/services/service"

# Name of the WMS street layer
# streets = "osm"
osm_lyr = "OSM-WMS"
```

```
# Name of the basemap image to save
osm_img = "basemap"

# OSM EPSG code (spatial reference system)
osm_epsg = 3857

# Optional WMS parameter
osm_style = ""
```

4. Next, we set up our hillshade parameters, which will determine the angle and direction of our artificial sun:

```
# Shaded elevation parameters
#
# Sun direction
azimuth = 315.0

# Sun angle
altitude = 45.0

# Elevation exageration
z = 5.0

# Resolution
scale = 1.0
```

5. Then we set up the `no_data` value where there is no elevation information:

```
# No data value for output
no_data = 0
```

6. Next, we set up the name of our output image as follows:

```
# Output elevation image name
elv_img = "elevation"
```

7. Now we create the colors for our minimum and maximum elevation values with the following code:

```
# RGBA color of the SRTM minimum elevation
min_clr = (255, 255, 255, 0)

# RGBA color of the SRTM maximum elevation
max_clr = (0, 0, 0, 0)

# No data color
zero_clr = (255, 255, 255, 255)
```

8. Then we set up our output image size, as follows:

```
# Pixel width and height of the

# output images
w = 800
h = 800
```

Now that we understand how the functions work, let's parse the GPX.

Parsing the GPX

Now, we'll parse the GPX file, which is just XML, using the `built-in`
`xml.dom.minidom` module. We'll extract latitude, longitude, elevation, and timestamps.
We'll store them in a list for later use. The timestamps are converted to `struct_time`
objects using Python's `time` module, which makes it easier to work with.

The following steps need to be performed for parsing:

1. First, we parse the `gpx` file using the `minidom` module:

```
# Parse the gpx file and extract the coordinates
log.info("Parsing GPX file: {}".format(gpx))
xml = minidom.parse(gpx)
```

2. Next, we get all of the `"trkpt"` tags that contain the elevation information:

```
# Grab all of the "trkpt" elements
trkpts = xml.getElementsByTagName("trkpt")
```

3. Now, we set up the lists to store our parsed location and elevation values:

```
# Latitude list
lats = []
# Longitude list
lons = []
# Elevation list
elvs = []
# GPX timestamp list
times = []
```

4. Then, we loop through the GPS entries in the GPX and parse the values:

```
# Parse lat/long, elevation and times
for trkpt in trkpts:
  # Latitude
```

```
lat = float(trkpt.attributes["lat"].value)
# Longitude
lon = float(trkpt.attributes["lon"].value)
lats.append(lat)
lons.append(lon)
# Elevation
elv = trkpt.childNodes[0].firstChild.nodeValue
elv = float(elv)
elvs.append(elv)
```

The timestamp requires a little bit of extra work because we have to convert from GMT time to local time:

```
# Times
t = trkpt.childNodes[1].firstChild.nodeValue
# Convert to local time epoch seconds
t = get_local_time(t)
times.append(t)
```

After we parse the GPX, we need the bounding box of the route to download data from other geospatial services.

Getting the bounding box

When we download data, we want the dataset to cover more area than the route so the map is not cropped too closely around the edges of the route. So we'll buffer the bounding box by 20% on each side. Finally, we'll need the data in Eastings and Northings to work with the WMS service. Eastings and Northings are the x and y coordinates of points in the Cartesian coordinate system in meters. They are commonly used in the UTM coordinate system:

1. First, we get the extents from our coordinate lists as follows:

```
# Find Lat/Long bounding box of the route
minx = min(lons)
miny = min(lats)
maxx = max(lons)
maxy = max(lats)
```

2. Next, we buffer the bounding box to ensure the track isn't taken close to the edge:

```
# Buffer the GPX bounding box by 20%
# so the track isn't too close to
# the edge of the image.
xdist = maxx - minx
ydist = maxy - miny
```

```
x20 = xdist * .2
y20 = ydist * .2

# 10% expansion on each side
minx -= x20
miny -= y20
maxx += x20
maxy += y20
```

3. Finally, we set up our bounding box in a variable and convert our coordinates to meters, which the web service requires:

```
# Store the bounding box in a single
# variable to streamline function calls
bbox = [minx, miny, maxx, maxy]

# We need the bounding box in meters
# for the OSM WMS service. We will
# download it in degrees though to
# match the SRTM file. The WMS spec
# says the input SRS should match the
# output but this custom service just
# doesn't work that way
mminx, mminy = ll2m(miny, minx)
mmaxx, mmaxy = ll2m(maxy, maxx)
```

With this, we will now download our map and elevation images.

Downloading map and elevation images

We'll download the OSM basemap first as our basemap, which has streets and labels:

1. First, we'll download the OSM basemap using `log.info`:

```
# Download the OSM basemap
log.info("Downloading basemap")
wms(mminx, mminy, mmaxx, mmaxy, osm_WMS, osm_lyr,
 osm_epsg, osm_style, osm_img, w, h)
```

This section will produce an intermediate image as shown in the following screenshot:

2. Next, we'll download some elevation data from the **SRTM** dataset. SRTM is nearly global and provides a 30-90 m resolution. The SRTM.py Python module makes working with this data easy. SRTM.py downloads the data and sets it needs to make a request. Therefore, if you download data from different areas, you may need to clean out the cache located in your home directory (~/.srtm). This part of the script can also take up to 2-3 minutes to complete, depending on your computer and internet connection speeds:

```
# Download the SRTM image

# srtm.py downloader
log.info("Retrieving SRTM elevation data")
# The SRTM module will try to use a local cache

# first and if needed download it.
srt = srtm.get_data()
# Get the image and return a PIL Image object
image = srt.get_image((w, h), (miny, maxy), (minx, maxx),
 300, zero_color=zero_clr, min_color=min_clr,
 max_color=max_clr)
# Save the image
image.save(elv_img + ".png")
```

This portion of the script also outputs an intermediate elevation image, as shown in the following screenshot:

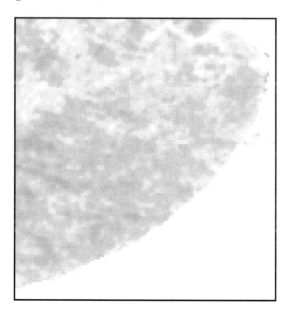

Now that we have our elevation image, we can turn it into a hillshade.

Creating the hillshade

We can run this data through the same **hillshade** algorithm used in *Creating a shaded-relief* section in Chapter 7, *Python and Elevation Data*. For this, let's follow these steps:

1. First, we open our elevation image and read it into a numpy array:

```
# Hillshade the elevation image
log.info("Hillshading elevation data")
im = Image.open(elv_img + ".png").convert("L")
dem = np.asarray(im)
```

2. Now we set up our processing windows to move through the grid and analyze it in small sections for efficiency:

```
# Set up structure for a 3x3 windows to
# process the slope throughout the grid
window = []
# x, y resolutions
xres = (maxx-minx)/w
yres = (maxy-miny)/h
```

3. Then, we break the elevation image into windows as follows:

```
# Create the windows
for row in range(3):
 for col in range(3):
 window.append(dem[row:(row + dem.shape[0]-2),
 col:(col + dem.shape[1]-2)])
```

4. We will create arrays for our processing windows as follows:

```
# Process each cell
x = ((z * window[0] + z * window[3] + z * window[3] + z *
window[6]) -
 (z * window[2] + z * window[5] + z * window[5] + z * window[8])) \
 / (8.0 * xres * scale)

y = ((z * window[6] + z * window[7] + z * window[7] + z *
window[8]) -
 (z * window[0] + z * window[1] + z * window[1] + z * window[2])) \
 / (8.0 * yres * scale)
```

5. Finally, we can process them in a single pass thanks to `numpy`:

```
# Calculate slope
slope = 90.0 - np.arctan(np.sqrt(x*x + y*y)) * rad2deg

# Calculate aspect
aspect = np.arctan2(x, y)

# Calculate the shaded relief
shaded = np.sin(altitude * deg2rad) * np.sin(slope * deg2rad) \
 + np.cos(altitude * deg2rad) * np.cos(slope * deg2rad) \
 * np.cos((azimuth - 90.0) * deg2rad - aspect)

shaded = shaded * 255
```

Now that we have our hillshade layer, we can begin creating maps.

Creating maps

We have the data we need to begin building the map for our report. Our approach will be the following:

- Enhancing the elevation and basemap images with filters
- Blending the images together to provide a hillshaded OSM map
- Creating a translucent layer to draw the street route
- Blending the route layer with the hillshaded map

These tasks will all be accomplished using the PIL `Image` and `ImageDraw` modules, as shown in the following steps:

1. First, we convert our shaded relief `numpy` array back to an image and smooth it:

```
# Convert the numpy array back to an image
relief = Image.fromarray(shaded).convert("L")

# Smooth the image several times so it's not pixelated
for i in range(10):
 relief = relief.filter(ImageFilter.SMOOTH_MORE)

log.info("Creating map image")
```

2. Now we'll increase the contrast in the image to make it stand out more:

```
# Increase the hillshade contrast to make
# it stand out more
e = ImageEnhance.Contrast(relief)
relief = e.enhance(2)
```

3. Next, we crop the map image to the same size as our elevation image:

```
# Crop the image to match the SRTM image. We lose
# 2 pixels during the hillshade process
base = Image.open(osm_img + ".jpg").crop((0, 0, w-2, h-2))
```

4. Then we increase the contrast on the map image as well and blend it with the hillshade image:

```
# Enhance basemap contrast before blending
e = ImageEnhance.Contrast(base)
base = e.enhance(1)

# Blend the the map and hillshade at 90% opacity
topo = Image.blend(relief.convert("RGB"), base, .9)
```

5. Now we're ready to draw the GPS tracks on our blended map by first converting our points to pixels:

```
# Draw the GPX tracks
# Convert the coordinates to pixels
points = []
for x, y in zip(lons, lats):
 px, py = world2pixel(x, y, w, h, bbox)
 points.append((px, py))
```

6. We also need to subtract the buffer from the edge buffer from the tracks image we are about to create:

```
# Crop the image size values to match the map
w -= 2
h -= 2
```

7. Next, we create a transparent image and draw our track as a red line:

```
# Set up a translucent image to draw the route.
# This technique allows us to see the streets
# and street names under the route line.

track = Image.new('RGBA', (w, h))
```

```
track_draw = ImageDraw.Draw(track)

# Route line will be red at 50% transparency (255/2=127)
track_draw.line(points, fill=(255, 0, 0, 127), width=4)
```

8. Now we can paste the track on our image with the following code:

```
# Paste onto the basemap using the drawing layer itself
# as a mask.
topo.paste(track, mask=track)
```

9. Now we'll draw a starting point on the route like so:

```
# Now we'll draw start and end points directly on top
# of our map - no need for transparency
topo_draw = ImageDraw.Draw(topo)

# Starting circle
start_lon, start_lat = (lons[0], lats[0])
start_x, start_y = world2pixel(start_lon, start_lat, w, h, bbox)
start_point = [start_x-10, start_y-10, start_x+10, start_y+10]
topo_draw.ellipse(start_point, fill="lightgreen", outline="black")
start_marker = [start_x-4, start_y-4, start_x+4, start_y+4]
topo_draw.ellipse(start_marker, fill="black", outline="white")
```

10. Following is the code snippet for the ending point:

```
# Ending circle
end_lon, end_lat = (lons[-1], lats[-1])
end_x, end_y = world2pixel(end_lon, end_lat, w, h, bbox)
end_point = [end_x-10, end_y-10, end_x+10, end_y+10]
topo_draw.ellipse(end_point, fill="red", outline="black")
end_marker = [end_x-4, end_y-4, end_x+4, end_y+4]
topo_draw.ellipse(end_marker, fill="black", outline="white")
```

Now that we have our track drawn, we're ready to place our geotagged photo.

Locating the photo

We'll use a photo taken with a cell phone that adds GPS location coordinates. You can download it from:

`https://raw.githubusercontent.com/GeospatialPython/Learn/master/RoutePhoto.jpg`.

Place the image in a directory named photos at the same level as the script. We'll only use one photo but the script can handle as man images as you want. We'll draw and place a photo icon on the map and then save the completed basemap, as shown in the following steps:

1. First, we get a list of images with the following code:

```
# Photo icon
images = glob.glob("photos/*.jpg")
```

2. Next, we loop through each image and grab its GPS information:

```
for i in images:
  e = exif(i)
```

3. Then, we parse the location info using our GPS function as follows:

```
photo_lat, photo_lon = gps(e)
#photo_lat, photo_lon = 30.311364, -89.324786
```

4. Now, we can convert the photo coordinates to image pixel coordinates:

```
photo_x, photo_y = world2pixel(photo_lon, photo_lat, w, h, bbox)
```

5. Then we'll draw an icon for the location of the photo with the following code:

```
topo_draw.rectangle([photo_x - 12, photo_y - 10, photo_x + 12, \
photo_y + 10], fill="black", outline="black")
topo_draw.rectangle([photo_x - 9, photo_y - 8, photo_x + 9, \
photo_y + 8], fill="white", outline="white")
topo_draw.polygon([(photo_x-8,photo_y+7), (photo_x-3,photo_y-1),
(photo_x+2,photo_y+7)], fill = "black")
topo_draw.polygon([(photo_x+2,photo_y+7), (photo_x+7,photo_y+3),
(photo_x+8,photo_y+7)], fill = "black")
```

6. And finally, we'll save our map like so:

```
# Save the topo map
topo.save("{}_topo.jpg".format(osm_img))
```

While not saved to the filesystem, the hillshaded elevation looks like the following:

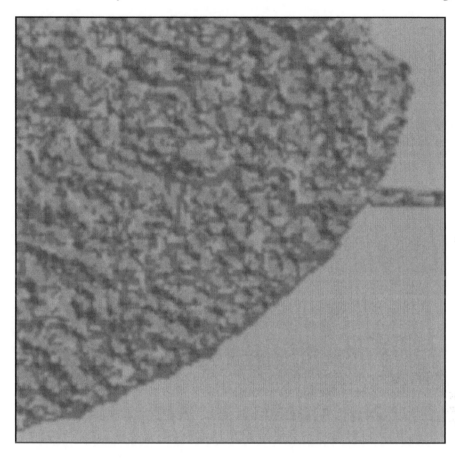

The blended topographic map looks like the following screenshot:

While hillshade mapping gives us an idea of the elevation, it doesn't give us any quantitative data. To get more detailed, we'll create a simple elevation chart.

Measuring elevation

Using the excellent Google Chart API, we can quickly build a nice elevation profile chart showing how the elevation changes across the route:

1. First, we'll create the `chart` object for our elevation profile:

```
# Build the elevation chart using the Google Charts API
log.info("Creating elevation profile chart")
chart = SimpleLineChart(600, 300, y_range=[min(elvs), max(elvs)])
```

2. Now, we need to create a line for our minimum value like so:

```
# API quirk - you need 3 lines of data to color
# in the plot so we add a line at the minimum value
# twice.
chart.add_data([min(elvs)]*2)
chart.add_data(elvs)
chart.add_data([min(elvs)]*2)

# Black lines
chart.set_colours(['000000'])
```

3. Next, we can fill in our elevation profile as follows:

```
# fill in the elevation area with a hex color
chart.add_fill_range('80C65A', 1, 2)
```

4. Then we can set up the elevation labels as follows and assign them to an axis:

```
# Set up labels for the minimum elevation, halfway value, and max
value
elv_labels = int(round(min(elvs))), int(min(elvs)+((max(elvs)-
min(elvs))/2))

# Assign the labels to an axis
elv_label = chart.set_axis_labels(Axis.LEFT, elv_labels)
```

5. Next, we can label the axis itself with the following code:

```
# Label the axis
elv_text = chart.set_axis_labels(Axis.LEFT, ["FEET"])
# Place the label at 30% the distance of the line
chart.set_axis_positions(elv_text, [30])
```

6. Now we can calculate the distance between the track points:

```
# Calculate distances between track segments
distances = []
measurements = []
coords = list(zip(lons, lats))
for i in range(len(coords)-1):
 x1, y1 = coords[i]
 x2, y2 = coords[i+1]
 d = haversine(x1, y1, x2, y2)
 distances.append(d)
total = sum(distances)
distances.append(0)
j = -1
```

We have the elevation profile, but we need to add the distance markers along the *x* axis so we know where along the route the profile changed.

Measuring distance

In order to understand the elevation data chart, we need reference points along the *x* axis to help us determine the elevation along the route. We will calculate the mile splits along the route and place those at the appropriate location on the x axis of our charts. Let's have a look at the following steps:

1. First, we locate the mile markers along the axis as follows:

```
# Locate the mile markers
for i in range(1, int(round(total))):
 mile = 0
 while mile < i:
 j += 1
 mile += distances[j]
 measurements.append((int(mile), j))
 j = -1
```

2. Next, we set up labels for the mile markers:

```
# Set up labels for the mile points

positions = []
miles = []
for m, i in measurements:
 pos = ((i*1.0)/len(elvs)) * 100
 positions.append(pos)
 miles.append(m)
```

```
# Position the mile marker labels along the x axis
miles_label = chart.set_axis_labels(Axis.BOTTOM, miles)
chart.set_axis_positions(miles_label, positions)
```

3. Now we can label the mile markers as follows:

```
# Label the x axis as "Miles"
miles_text = chart.set_axis_labels(Axis.BOTTOM, ["MILES", ])
chart.set_axis_positions(miles_text, [50, ])

# Save the chart
chart.download('{}_profile.png'.format(elv_img))
```

Our chart should now look like the following screenshot:

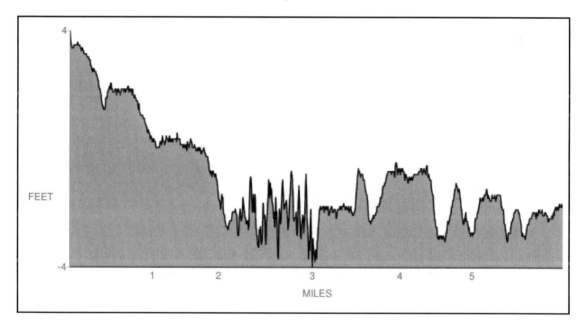

Our first chart is complete. Now, let's look at weather data along the route.

Retrieving weather data

In this section, we will retrieve our final data element: the weather. As mentioned earlier, we will use the Dark Sky service, which allows us to gather historical weather reports for any place in the world. The weather API is REST and JSON-based, so we'll use the `urllib` module to request data and the `json` library to parse it. Of note in this section is that we cache the data locally, so you can run the script offline for testing if need be. Early on in this section is where you place your Dark Sky API key that is flagged by the YOUR KEY HERE text. Let's have a look at the following steps:

1. First, we need the center of our area of interest:

```
log.info("Creating weather summary")

# Get the bounding box centroid for georeferencing weather data
centx = minx + ((maxx-minx)/2)
centy = miny + ((maxy-miny)/2)
```

2. Now, we set up the free Dark API key as follows so we can retrieve weather data:

```
# DarkSky API key
# You must register for free at DarkSky.net
# to get a key to insert here.
api_key = "YOUR API KEY GOES HERE"
```

3. Then, we grab the latest timestamp from our data that we'll use for our weather query:

```
# Grab the latest route time stamp to query weather history
t = times[-1]
```

4. Now we're ready to do our weather data query as follows:

```
history_req =
"https://api.darksky.net/forecast/{}/".format(api_key)
#name_info = [t.tm_year, t.tm_mon, t.tm_mday,
route_url.split(".")[0]]
#history_req += "/history_{0}{1:02d}{2:02d}/q/{3}.json"
.format(*name_info)
history_req += "{},{},{}".format(centy, centx, t)
history_req += "?exclude=currently,minutely,hourly,alerts,flags"
request = urllib.request.urlopen(history_req)
weather_data = request.read()
```

5. We'll cache the weather data like so just in case we want to look at it later:

```
# Cache weather data for testing
with open("weather.json", "w") as f:
  f.write(weather_data.decode("utf-8"))
```

6. Then we parse the weather JSON data as follows:

```
# Retrieve weather data
js = json.loads(open("weather.json").read())
history = js["daily"]
```

7. All we need is the weather summary, which is the first item in the list:

```
# Grab the weather summary data.
# First item in a list.
daily = history["data"][0]
```

8. Now, we'll get the specific weather attributes as follows:

```
# Max temperature in Imperial units (Farenheit).
# Celsius would be metric: "maxtempm"
maxtemp = daily["temperatureMax"]

# Minimum temperature
mintemp = daily["temperatureMin"]

# Maximum humidity
maxhum = daily["humidity"]

# Precipitation in inches (cm = precipm)
if "precipAccumulation" in daily:
 precip = daily["precipAccumulation"]
else:
 precip = "0.0"
```

9. Now that we have the weather data stored in variables, we can complete the final step: adding it all to a PDF report.

The fpdf library has no dependencies except PIL in some cases. For our purposes, it will work quite well. We are going to proceed down the page and add the elements. fpdf.ln() separates rows, while fpdf.cells contains text and allows for more precise layouts.

We're finally ready to create our PDF report with the following steps:

1. First, we set up our pdf object as follows:

```
# Simple fpdf.py library for our report.
# New pdf, portrait mode, inches, letter size
# (8.5 in. x 11 in.)
pdf = fpdf.FPDF("P", "in", "Letter")
```

2. Then, we'll add a page for our report and set our font preferences:

```
# Add our one report page
pdf.add_page()

# Set up the title
pdf.set_font('Arial', 'B', 20)
```

3. We'll create a title for our report with the following code:

```
# Cells contain text or space items horizontally
pdf.cell(6.25, 1, 'GPX Report', border=0, align="C")

# Lines space items vertically (units are in inches)
pdf.ln(h=1)
pdf.cell(1.75)

# Create a horizontal rule line
pdf.cell(4, border="T")
pdf.ln(h=0)
pdf.set_font('Arial', style='B', size=14)
```

4. Now, we can add the route map like so:

```
# Set up the route map
pdf.cell(w=1.2, h=1, txt="Route Map", border=0, align="C")
pdf.image("{}_topo.jpg".format(osm_img), 1, 2, 4, 4)
pdf.ln(h=4.35)
```

5. Next, we add the elevation chart as follows:

```
# Add the elevation chart
pdf.set_font('Arial', style='B', size=14)
pdf.cell(w=1.2, h=1, txt="Elevation Profile", border=0, align="C")
pdf.image("{}_profile.png".format(elv_img), 1, 6.5, 4, 2)
pdf.ln(h=2.4)
```

6. Then we can write the weather data summary with the following code:

```
# Write the weather summary
pdf.set_font('Arial', style='B', size=14)
pdf.cell(1.2, 1, "Weather Summary", align="C")
pdf.ln(h=.25)
pdf.set_font('Arial', style='', size=12)
pdf.cell(1.8, 1, "Min. Temp.: {}".format(mintemp), align="L")
pdf.cell(1.2, 1, "Max. Hum.: {}".format(maxhum), align="L")
pdf.ln(h=.25)
pdf.cell(1.8, 1, "Max. Temp.: {}".format(maxtemp), align="L")
pdf.cell(1.2, 1, "Precip.: {}".format(precip), align="L")
pdf.ln(h=.25)
```

7. The Dark Sky terms require us to add a logo to our report, crediting the excellent data source:

```
# Give Dark Sky credit for a great service (https://git.io/fjwHl)
pdf.image("darksky.png", 3.3, 9, 1.75, .25)
```

8. Now we can add the geolocated image with the following code:

```
# Add the images for any geolocated photos
pdf.ln(h=2.4)
pdf.set_font('Arial', style='B', size=14)
pdf.cell(1.2, 1, "Photos", align="C")
pdf.ln(h=.25)
for i in images:
  pdf.image(i, 1.2, 1, 3, 3)
  pdf.ln(h=.25)
```

9. And finally, we can save the report and view it:

```
# Save the report
log.info("Saving report pdf")
pdf.output('report.pdf', 'F')
```

You should have a PDF document in your working directory called
`report.pdf` containing your finished product. It should look like the image shown in the
following screenshot:

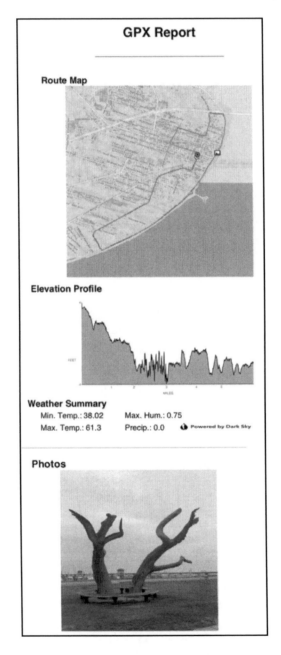

With this, we have used all the techniqueswe've learned throughout this book and built a GPS reporting tool.

Summary

Congratulations! In this book, you pulled together the most essential tools and skills needed to be a modern geospatial analyst. Whether you use geospatial data occasionally or use it all the time, you will be better equipped to make the most of geospatial analysis. This book focuses on using open source tools almost entirely found within the PyPI directory for ease of installation and integration. But even if you are using Python as a driver for a commercial GIS package or a popular library such as GDAL, the ability to test out new concepts in pure Python will always come in handy.

Further reading

Python provides a rich set of libraries for visualizing data. One of the most prominent is **Matplotlib**, which can produce numerous types of charts and maps and save them to PDF. Packt has several books on Matplotlib, including the *Matplotlib 30 Cookbook*: `https://www.packtpub.com/big-data-and-business-intelligence/matplotlib-30-cookbook`.

Other Books You May Enjoy

If you enjoyed this book, you may be interested in these other books by Packt:

Mastering Geospatial Development with QGIS 3.x - Third Edition
John Van Hoesen, GISP, Luigi Pirelli, Et al

ISBN: 978-1-78899-989-2

- Create and manage a spatial database
- Get to know advanced techniques to style GIS data
- Prepare both vector and raster data for processing
- Add heat maps, live layer effects, and labels to your maps
- Master LAStools and GRASS integration with the Processing Toolbox
- Edit and repair topological data errors
- Automate workflows with batch processing and the QGIS Graphical Modeler
- Integrate Python scripting into your data processing workflows
- Develop your own QGIS plugins

Learn QGIS - Fourth Edition
Anita Graser, Andrew Cutts

ISBN: 978-1-78899-742-3

- Explore various ways to load data into QGIS
- Understand how to style data and present it in a map
- Create maps and explore ways to expand them
- Get acquainted with the new processing toolbox in QGIS 3.4
- Manipulate your geospatial data and gain quality insights
- Understand how to customize QGIS 3.4
- Work with QGIS 3.4 in 3D

Leave a review - let other readers know what you think

Please share your thoughts on this book with others by leaving a review on the site that you bought it from. If you purchased the book from Amazon, please leave us an honest review on this book's Amazon page. This is vital so that other potential readers can see and use your unbiased opinion to make purchasing decisions, we can understand what our customers think about our products, and our authors can see your feedback on the title that they have worked with Packt to create. It will only take a few minutes of your time, but is valuable to other potential customers, our authors, and Packt. Thank you!

Index

Printed in Great Britain
by Amazon